Medicine and Biomedical Sciences in Modern History

Series Editors
Carsten Timmermann
University of Manchester
Manchester, UK

Michael Worboys
University of Manchester
Manchester, UK

The aim of this series is to illuminate the development and impact of medicine and the biomedical sciences in the modern era. The series was founded by the late Professor John Pickstone, and its ambitions reflect his commitment to the integrated study of medicine, science and technology in their contexts. He repeatedly commented that it was a pity that the foundation discipline of the field, for which he popularized the acronym 'HSTM' (History of Science, Technology and Medicine) had been the history of science rather than the history of medicine. His point was that historians of science had too often focused just on scientific ideas and institutions, while historians of medicine always had to consider the understanding, management and meanings of diseases in their socio-economic, cultural, technological and political contexts. In the event, most of the books in the series dealt with medicine and the biomedical sciences, and the changed series title reflects this. However, as the new editors we share Professor Pickstone's enthusiasm for the integrated study of medicine, science and technology, encouraging studies on biomedical science, translational medicine, clinical practice, disease histories, medical technologies, medical specialisms and health policies.

The books in this series will present medicine and biomedical science as crucial features of modern culture, analysing their economic, social and political aspects, while not neglecting their expert content and context. Our authors investigate the uses and consequences of technical knowledge, and how it shaped, and was shaped by, particular economic, social and political structures. In re-launching the Series, we hope to build on its strengths but extend its geographical range beyond Western Europe and North America.

Medicine and Biomedical Sciences in Modern History is intended to supply analysis and stimulate debate. All books are based on searching historical study of topics which are important, not least because they cut across conventional academic boundaries. They should appeal not just to historians, nor just to medical practitioners, scientists and engineers, but to all who are interested in the place of medicine and biomedical sciences in modern history.

This series continues the Science, Technology and Medicine in Modern History series.

Mauro Capocci • Daniele Cozzoli
Editors

Empire, Nation-building, and the Age of Tropical Medicine, 1885–1960

palgrave
macmillan

Editors
Mauro Capocci
Department of Civilizations and Forms of Knowledge
University of Pisa
Pisa, Italy

Daniele Cozzoli
Department of Humanities
Pompeu Fabra University
Barcelona, Spain

ISSN 2947-9142 ISSN 2947-9150 (electronic)
Medicine and Biomedical Sciences in Modern History
ISBN 978-3-031-38804-0 ISBN 978-3-031-38805-7 (eBook)
https://doi.org/10.1007/978-3-031-38805-7

This Palgrave Macmillan imprint is published by the registered company Springer Nature Switzerland AG.
The registered company address is: Gewerbestrasse 11, 6330 Cham, Switzerland

To Valentina, Elena, Marta: everything everywhere all at once
and to Giulia, in memoriam
ognuno sta solo sul cuor della terra
trafitto da un raggio di sole:
ed è subito sera.

Acknowledgments

Most of the chapters of the book were presented at the workshop on Empires, the Nation, and Tropical Medicine (1885–1960), which took place in Barcelona from 22 to 23 November 2018, financed by a Spanish Ministry of Innovation and Competitiveness grant (ref. HAR2016-75559-P), whose PI were Mariciό Janué and Albert Presas i Puig and by the Department of Humanities, UPF, and the University Institute of History Jaume Vicens Vives, UPF. The editors wish to thank all the participants. We also would like to thank Claudia Contente and Josep Maria Fradera, who read a draft of the introductive essay, for their insightful comments.

Contents

1 **Introduction: The Age of Empire, the Making of the Modern Nation, and the Advancement of Medical Sciences** 1
Mauro Capocci and Daniele Cozzoli

Part I Tropical Medicine in the Evolution and the Collapse of Empires 27

2 **Tropical Medicine and the "Consolidation" of the Portuguese Empire (1902–1966)** 29
Isabel Amaral

3 **Dutch Colonial Medicine and Empire Building in the Tropics: The Cases of Leprosy and Drug Use in the Dutch East and West Indies Compared** 53
Stephen Snelders

Part II Tropical Medical Institutions and Imperial, Commercial, and Political Expansion 73

4 The Business of Tropical Medicine: Connections Between Anti-malarial Campaigns in Sierra Leone, 1899–1901, and Jamaica, 1908 75
Juanita De Barros and Deborah J. Neill

5 Leishmaniases in Brazil: A Historical Approach 101
Jaime Larry Benchimol

Part III Circulation of People, Object, and Ideas 137

6 Tropical Medicine, the Nation, and the Colonial Expansion in the View of Italian Royal Navy Physicians at the End of the Nineteenth Century 139
Mauro Capocci and Daniele Cozzoli

7 From Universal Rats to Future Jungle Foci: Actors and Places of Plague in Brazil (1899–1940s) 161
Matheus Alves Duarte da Silva

8 Anti-Fascist Medicine and the International Peace Campaign Against Urban Raids in Spain and China, 1936–1939 181
Carles Brasó Broggi

Correction to: From Universal Rats to Future Jungle Foci: Actors and Places of Plague in Brazil (1899–1940s) C1
Matheus Alves Duarte da Silva

Index 207

Notes on Contributors

Isabel Amaral is an Associate Professor of Habilitation at NOVA School of Science and Engineering. Since 2020, she is the co-coordinator of the Interuniversity Center for the History of Science and Technology (CIUHCT). Additionally, she is a member of the Council of R&D Units at the NOVA School of Science and Engineering, where she oversees the scientific management of researchers within the institution.

Isabel Amaral holds positions as a member of the Scientific Council and the Ethical Council at NOVA School of Sciences and Technology. She also serves on the Editorial Board of *Medical History*, Cambridge Core, as well as the *HoST—Journal of the History of Science*. Her research interests primarily lie in the history of the biological and biomedical sciences (nineteenth to twenty-first centuries), scientific and medical heritage, public health policies, and bioethics. She has published extensively and undertaken research projects focussed on these fields, with a particular emphasis on utilizing the Portuguese case as a case study.

Juanita De Barros is a professor in the Department of History at McMaster University and the director of McMaster University's Centre for Human Rights and Restorative Justice. Her research concentrates on the history of health, gender, and reproductive rights in the Caribbean within the context of imperialism and post-slavery societies. Her most recent books are *Reproducing the British Caribbean: Sex,* Gender, *and Population Politics After Slavery* (2014) and the co-edited essay collection, *Public Health and the Imperial Project* (2015).

Jaime Larry Benchimol, Casa de Oswaldo Cruz/Fundação Oswaldo Cruz, Rio de Janeiro, Brazil, holds a doctorate in history from the Universidade Federal Fluminense (1995). He was science editor of *História, Ciências, Saúde—Manguinhos* from 1997 to 2015 and continues to serve on the journal's Editorial Board. He is currently a senior researcher at the Casa de Oswaldo Cruz, Oswaldo Cruz Foundation, and a professor in the Graduate Program in the History of Science and Health at the Casa de Oswaldo Cruz. His research and teaching focus chiefly on the history of the life sciences and the history of tropical medicine and public health. His current work concentrates on the history of the leishmaniases in the New World.

Carles Brasó Broggi, PhD in history, Pompeu Fabra University, Barcelona, 2010. He is a sociologist and economic historian. He is Ramón y Cajal fellow (2020–2025) at Universitat Oberta de Catalunya and a member of ALTER research group. His investigations deal with knowledge and technology transfers between Europe and China in the twentieth century. He has published an economic history book about China's textile industry (*Trade and Technology Networks in the Chinese Textile Industry*, Palgrave Macmillan, 2016) and a global history monograph on internationalist doctors between Spain and China (*Los médicos errantes*, Crítica, 2022).

Mauro Capocci is an Associate Professor of History of Science and Medicine at the University of Pisa, Italy. A graduate in philosophy at Sapienza University of Rome, he was awarded a PhD at the University of Florence. His research focusses mostly on the history of contemporary Italian biomedical sciences, as detailed in several publications in international journals. A member of the European Society for the History of Science and other Italian academic Societies, Mauro is in the Editorial Board of the journals *Medicina nei Secoli* -, *Journal of History of Medicine and Medical Humanities,* and *Physis*. He also contributes regularly to *Le Scienze*, the Italian edition of Scientific American.

Daniele Cozzoli, PhD, University of Rome "La Sapienza," is an Associate Professor (Professor Agregat) of the History of Science at Pompeu Fabra University, Barcelona. He is editor in chief of the journal *Centaurus*, a correspondent member of the International Academy of the History of Science, a member of the international board of the Society for the History of Humanities, and an advisory board member of the journal *Medical*

History. Daniele has published two books in Italian and a number of articles and chapters on the history of science in the sixteenth and seventeenth centuries, biomedical sciences in the twentieth century, and historiography of science.

Deborah J. Neill is an associate professor in the Department of History at York University, Toronto. A European historian with an interest in imperialism and empire in the late nineteenth and early twentieth centuries, her research focuses on medicine and health, food and nutrition, and economic and business history. She is the author of *Networks in Tropical Medicine: Internationalism, Colonialism, and the Rise of a Medical Specialty, 1890–1930* (2012).

Matheus Alves Duarte da Silva is a postdoctoral research fellow at the University of St Andrews, UK, working on the global history of medicine, zoonoses, and disease ecology. He is the author of *Quand la peste connectait le monde: production et circulation de savoirs microbiologiques entre Brésil, Inde et France (1894–1922)* (EHESS, 2020), and co-edited the book *Beyond Science and Empire: Circulation of Knowledge in an Age of Global Empires 1750–1945* (2023). He has authored several chapters and articles on the global history of plague, the history of microbiology, and the history of science in Latin America.

Stephen Snelders studied History at Utrecht University, the Netherlands, and obtained his PhD in 1999 at VU-University, Amsterdam. Since 2015, he works as a researcher and lecturer in the History and Philosophy of Science-section of the Freudenthal Institute, Faculty of Science, Utrecht University. His research interests include the history of drugs and of colonial and tropical medicine. He published books on *Leprosy and Colonialism: Suriname Under Dutch Rule, 1750–1950* (2017) and *Drug Smuggler Nation: Narcotics and* the Netherlands, *1920–1995* (2021, 2023).

List of Figures

Fig. 2.1 The parasitology class of Ayres Kopke at the School of Tropical Medicine. IHMT 0000350 39

Fig. 2.2 Brigade of servicemen with suits for capturing tsetse flies (of the *glossina* genus) on Príncipe Island. Relatório final da missão da doença do sono na ilha do Príncipe. Arquivos de Higiene e Patologia Exóticas, 1915: 111 43

Fig. 2.3 Representation of the scientific mission of sleeping sickness in Guinea by a Guinean artist in 2014. (Courtesy of Luis Neves da Costa) 49

CHAPTER 1

Introduction: The Age of Empire, the Making of the Modern Nation, and the Advancement of Medical Sciences

Mauro Capocci and Daniele Cozzoli

The essays collected in this book deal with the complex relation between the development of modern empires and the history of tropical medicine. We aim at broadening existing historiographical perspectives by discussing a number of experiences that have been poorly considered hitherto. Most of the recent literature in English language has, indeed, focused on the history of tropical medicine in the British world system (Crozier (then Greenwood), 2007; Edmond, 2006; Pati & Harrison, 2009; Johnson, 2010; Greenwood & Topiwala, 2015; Greenwood, ed., 2016; Haynes,

M. Capocci
Department of Civilizations and Forms of Knowledge, University of Pisa, Pisa, Italy
e-mail: mauro.capocci@unipi.it

D. Cozzoli (✉)
Department of Humanities, Pompeu Fabra University, Barcelona, Spain
e-mail: daniele.cozzoli@upf.edu

M. Capocci, D. Cozzoli (eds.), *Empire, Nation-building, and the Age of Tropical Medicine, 1885–1960*, Medicine and Biomedical Sciences in Modern History,
https://doi.org/10.1007/978-3-031-38805-7_1

2017). However, according to Michael A. Osborne (2014), the explanatory model adopted for the British case cannot be easily extended to other colonial experiences. In the late nineteenth century, European and Asian countries created colonial and imperial dominions, and each of these 'systems'[1] adopted multifaceted approaches to colonial health and medicine. By means of a larger picture of empires, it is thus possible to obtain a broader image of what is now called tropical medicine. Our aim is not to reduce the complexity of the dynamics of the actors involved in the development of tropical medicine to its function in promoting colonial expansion. Rather, as historians of science and medicine, we aim to engage with the approach of recent literature on imperial history, in order to contribute to shed light on a number of facets both of the history of tropical medicine and of the contemporary processes of nation-building and empire-building (Elliott, 2018; Fradera, 2018; Paquette, 2020).

Christopher Bayly defined the period between 1780 and 1914 (the so-called Long Nineteenth century) as 'the birth of the modern world,' because it was in this period that the nation-state and a massive expansion of circulation of goods, people and ideas made the world more homogeneous (Bayly, 2004). Such processes continued throughout the twentieth century. The time span encompassed by this collection witnessed the development of the modern world, in terms of not only the expansion of European powers but also of the increase of population mobility from and to certain areas. Since the early decades of nineteenth century, changes in slavery legislation caused large migrations of Chinese, South Asian and European workers (see, for example, McNeill, 2010, p. 288) towards the Americas. Even larger movements took place at the turn of the century from Ireland and Italy (Hochstadt, 1999). The Berlin conference in 1884 opened the way for a new wave of colonial expansion, mostly directed towards Africa: in a few decades, a handful of Western powers either occupied or economically controlled the rest of the world. Formal empires grew in size and numbers. Germs travelled together with people and commodities: leprosy, cholera, plague and yellow fever are some of the scourges that threatened the newly interconnected world. With the increased mobility, emerging or re-emerging diseases—respectively, cholera and yellow fever, and plague and leprosy—called for action. The Pasteurian revolution eventually changed the medical framework and opened the way to new treatment and preventive measures. A new approach to sanitation and

[1] For the world-system analysis, see below.

hygiene, sera, vaccines and a small number of drugs entered the medical arsenal, opening new grounds for research and public health (Dubos, 1995; Geison, 1995; Gradmann, 2009). Nonetheless, the consequences of bacteriological discoveries were not immediate. During the Spanish-American War of 1895–1898, 13 out of every 14 casualties were still due to diseases (Cueto, 2008).

Historiography mostly refers to the dichotomy 'sanitation vs. contagionism,' following Ackerknecht's (1948) classical account. In the nineteenth century, after the demise of the plague and the blooming of the liberal colonial order of the British Empire, 'anti-contagionism' gained new traction. Contagionism was increasingly considered a relic of the past, a proxy for quarantine and other illiberal, unscientific and outdated measures that unnecessarily hindered the free circulation of people and goods. The cholera waves that hit Europe since the 1820s pushed such dichotomy deep into European politics, though the opinion of the medical community was often oscillating. However, the medical framework deployed to understand cholera was mostly based on two main features: class hierarchy and colonialism (Hamlin, 2009). Though with some resistance, in the last decade of the nineteenth century, Pasteur's and Koch's germ theory reinstated contagion as a scientific phenomenon and allowed to fight infectious diseases without recurring to quarantine. The conceptual expansion following Pasteurian innovation was also a geographical expansion: medical sciences found new fields in the 'warm climates,' and colonial medicine, in research and policy, often followed political and diplomatic fault-lines, within and among empires. For example, while the increased interconnectedness of the world meant a growing exchange of people, commodities and pathogens, the elites used medicine as one tool to control migration: sanitary restrictions were applied to 'coolies' from China in California or to Polish seasonal workers in East Prussia (Conrad, 2010). The growing population of Latin American cities and European ports encouraged the spread of epidemics. These cities became the focus of new efforts in controlling contagion at a global level, with medical officers and researchers converging from European countries to contribute.

Michael A. Osborne (2014) has done substantial work on the French tropical medicine, and in this book we aim to extend his perspective on a number of experiences in the history of tropical medicine during the Age of Empire, experiences connecting China with Spain, Italy with Africa and South America, Britain with Sierra Leone and Jamaica, the Netherlands with its American and Asian colonies, Portugal and its African colonies, or focusing on poorly studied countries such as Brazil. Under different conditions and in different roles, there were many actors on the global stage. Looking at Italy, for example, we see at least three different layers: 'tropical diseases' like malaria or leishmaniasis were in fact endemic in the country; there was a process of nation building going on, involving colonial expansion in different areas of Africa and the Mediterranean; massive migrations were taking place mainly towards North and South America. These three facets—from the medical point of view—meant that Italians were exposed to tropical diseases in different forms: as citizens, as empire builders, as working-class migrants. As a result, the development of tropical/colonial/imperial medicine in Italy was intertwined with national politics but interwoven in international networks.

Some countries experiencing epidemics were the focus of acute international competition: Rio de Janeiro suffered deadly epidemics of yellow fever, and the French government asked the Pasteur Institute to send a mission to Brazil in 1901 in order to research the disease. Though Brazil was outside the French imperial boundaries, the disease was not: it ravaged the French colony of Senegal, and Brazil offered the chance to study the pathology in a more favourable setting (Benchimol, 1999; Löwy, 1990, 2001).

Migrations were in fact one of the most relevant phenomena from the medical point of view, and biomedical researchers and practitioners deployed a vast array of tools to control the massive movement of people in the newly interconnected world. The Bertillon biometric system was not just for criminals but for foreigners too (Conrad, 2010; About, 2011; Galeano & García Ferrari, 2011). In 1881, the United States began pressuring other countries to check on health conditions of immigrants, even posting American physicians to European ports to integrate local examinations. In Germany, as of the 1890s, nationalist physicians and medical investigators were concerned that immigrants from Eastern Europe could also bring typhus, cholera and other infectious diseases (Weindling, 2000; Baldwin, 1999; Conrad, 2010).

A number of biomedical institutions played a crucial role in colonial policies. The Liverpool and London Schools of Tropical Medicine were primarily created to support the British expansion in West Africa (Johnson, 2010). The Pasteur Institute established affiliated institutes in the colonies and in areas where Paris had a strategic interest, but at the same time it promoted the image of Pasteur as a universal hero of mankind, contributing to construct the view of the French civilizing mission worldwide

(Nicolle, 1935; Delaunay, 1962; Moulin, 1992; Dedet, 2000; Cozzoli, 2018).

National pride and necessity were never foreign to the creation of policies against the spread of infectious disease within the international context.

We propose to integrate a local perspective or a perspective centred in the relation between a colony and its motherland with the interactions among empires and among nations, as the emerging picture of tropical medicine is extremely complex. This is the second aspect that this collection of essays explores. As already recalled, recent studies have primarily focused on how tropical medicine developed as part of colonial efforts in Britain (with the notable exception of Neill, 2012; Osborne, 2014) providing crucial insights in a number of facets of tropical medicine in the British world system (Harrison, 1999; Crozier, 2007; Johnson, 2010; Osborne, 2014, Haynes, 2017). Throughout the twentieth century, tropical medicine in Britain was deeply transformed by the encounter with the peoples of the Empire. Anna Crozier (Greenwood) and Ryan Johnson have focused on the colonial medical service in British East and West Africa (Crozier, 2007; Johnson, 2010; Greenwood, 2016; Greenwood & Topiwala, 2015), while Douglas M. Haynes has explained how women, Jewish refugees and colonial Indian doctors changed the white male nature of the medical profession in Britain (Haynes, 2017). All the abovementioned authors have dealt with tropical medicine in the British world system without working out the connections with the development of the discipline in other contemporary cases. Even if our perspective shifts the focus from Britain, it can be inspiring to students of the British tropical medicine too. Like Bhattacharya et al. (2005) for vaccination in India, Johnson (2010) has argued that, despite the imperial rhetoric, tropical medicine in British West Africa was not 'a tool of the empire.' However, the perspective changes if instead of looking at the British empire as 'an empire by design,' we follow John Darwin's (2009) view: empire was rather a 'world system,' an interconnected structure that emerged as a consequence of events and circumstances not always planned. Darwin's view is rooted in Braudel's idea of 'world-economy,' then extended by Wallerstein's 'world-system notion.' According to Braudel, determinate regions of the world form in certain epochs an integrated economic space, with slow-moving borders, a pivot city or area, and other areas organised into a hierarchy instrumental to the same world-economy. In a similar perspective, Wallerstein claimed that since the sixteenth century a world-system emerged in connection to the rise of capitalism (Wallerstein, 2004, p. 22; 2011, vol. 2). Within this world-system, in the mid-1850s, Britain became an 'hegemonic power,' a status defined by Wallerstein in the following way:

> What allows us to call them hegemonic is that for a certain period they were able to establish the rules of the game in the inter-state system, to dominate the world-economy (in production, commerce and finance), to get their way politically with a minimal use of military force (which however they had in good strength), and to formulate the cultural language with which one discussed the world. (Wallerstein 2004, pp. 57–58)

By assuming this perspective, 'Empire' stands for a wide and diverse range of interacting systems. We do not want to analyse what function tropical medicine had in the establishment of colonial domination, but rather how it emerged within a complex history that was also the result of the interaction between empires, metropolis and colonies. Empires were not closed entities. We can, for instance, observe how tropical medicine often unfolded within intertwined geopolitical global competition. These motives were often coupled by biomedical institutions and industrial companies in a non-linear way. The history of Bayer 205 (Suramin)—the first highly effective drug against sleeping sickness (African trypanosomiasis)—is a prime example. The chemical was pompously renamed Germanin and tested by a number of German researchers in the Belgian Congo and British Central Africa (Neill, 2012). Rumours spread that the Germans aimed to trade the Germanin formula for the return of their African colonies lost after WWI. On 25 August 1924, *The Times* reported that at the meeting of the Association for Tropical Diseases in Hamburg claims were made that Bayer 205 was the key to the restitution of German colonies, as Bayer kept the formula of the Germanin secret to prevent competitors from copying it. Such claims were echoed by British researchers and collaboration with Germany was discontinued. At the Pasteur Institute, Ernest Fourneau and his team succeeded in finding the Bayer 205 formula and Rhône-Poulenc (a French pharma company that supported Fourneau's laboratory) marketed an equivalent drug. However, it is highly implausible that the Germans really intended to trade the chemical formula of a drug for colonial gains, as sooner or later competitors would find it. It is more likely that British chemists and companies were interested in halting German field trials, while the Pasteur Institute/Rhône-Poulenc researchers aimed at presenting unfair competition as a patriotic act (Pope, 1924; typescript of Ernest Fourneau and letter from Louis Pasteur Vallery-Radot to the Minister of Internal Affairs of 29 September 1944, in Archives de l'Institut Pasteur/Laboratoire de Chimie Thérapéutique/1; Delaunay, 1962; Bovet, 1989). The Germans also relied on the Soviets and organized trials on Bayer trypanosomiasis drugs with them. During the 1920s, German biomedical researchers cooperated actively with the Soviets: a collaboration that was instrumental to regaining influence in Russia. In June 1922, under the umbrella of the German Red Cross, the Germans took over control of the former German Hospital in Petrograd (Weindling, 2000; Solomon Gross, 2006).

Furthermore, in the international institutions where the new globalization enacted its rituals—such as the Alexandria Quarantine and Health Board—the public health attitudes of the European delegates often played along with the best interest of the nation. In such settings, the abovementioned Ackernecht's dichotomy is only partially useful to understand the development of colonial medicine. As argued by Baldwin (1999), there was no clear-cut opposition between liberalism (anti-quarantine) and conservative authoritarianism (quarantine), but rather a shifting balance according to local and temporary needs. Case studies regarding different countries and backgrounds and fine-tooth combing of the documents are needed in this respect in order to avoid the oversimplification of complex global-scale interactions.

In the following sections, we will introduce some relevant aspects of Imperial and global history that clearly affected colonial and tropical medicine. First, we will discuss the relationship between imperialism(s) and nation-building, while later sections focus on colonialism (in a broadened perspective) and medicine. The last section will summarize the content of the collected essays.

The Complexity of Empires and the Making of the Nation

As the birthplace of the Industrial Revolution, Britain as of the mid-nineteenth century acquired a central position in the world economy, becoming the world's largest importer and exporter of commodities. Britons developed a number of pivotal technological innovations, such as the spinning jenny and the steam engine, while British fleets and financial services became essential to the world economy. Other countries specialized in satisfying the demands of British buyers: cotton from the Southern states of the United States until the American Civil War, guano from Peru, wool from Australia, wine from Portugal, nitrates and copper from Chile, beef and wheat from Argentina, dairy products from Denmark (Hobsbawm, 1996). In the eighteenth and nineteenth centuries, Britain was essential to the world. It comes, therefore, as no surprise that most historians of medicine have shaped their imperial and colonial narrative keeping in mind the development of the British Empire and its dramatic demise in the post-WWII years. Britain was essential to the world, but it also was unique.

In the timespan we address in this book, Britain domination was changing. The British Empire reached the acme of its complexity—including formal and informal colonies, protectorates, mandates, settlements, military fortresses, concessions, 'spheres of influence' and even a composite motherland, the United Kingdom of Great Britain and Ireland (until 1922). At the beginning of the twentieth century more than 40 juridical categories of territories existed, administered by three ministries—the General Office, the India Office and the Foreign Office. As already recalled, John Darwin prefers the expression 'British world system' to that of 'British Empire,' stressing that while imposing a rational structure on this expansion was beyond the power of the imperial government, nevertheless, an interconnected system emerged (Ferguson, 2004; Osterhammel, 1995; Darwin, 2009; Edgerton, 2019). In the long nineteenth century, Britain was the leading economic power and the centre of the world system, as the result of a series of unpredictable events and circumstances, including economic and political changes in Asia, patterns of consumption, religious factors and movements of people. Yet, the same factors determined the limits in space and time of British supremacy. The British economy was global, and changes in the global economy greatly affected it. By the early 1890s, Britain began to lose its industrial primacy. The United States and Germany surpassed British steel production. Germany (until 1914), the United States and Japan created informal empires as well (Hobsbawm, 1996; Osterhammel, 2010; Edgerton, 2019). British decadence became apparent with the Russian Revolution, as it started to be less essential to the world economic system. With the end of the Gold Standard in 1931, Britain ceased to be the economic hub of the world (Friedman & Schwartz, 1963; Eichengreen, 1992; Hobsbawm, 1996; Di Nolfo, 2007), and after WWII the decolonization process and the demise of the Commonwealth system of privileged exchanges finally marked the end of Britain's centrality in the world economy (Di Nolfo, 2015; Edgerton, 2019).

During the Age of Empire, the world was dominated by a number of formal and informal empires, inter-oceanic—seaborne in Gabriel Paquette's words—(British, French, Dutch, Portuguese, Spanish) or mostly continental (Russian, Austro-Hungarian, Ottoman, Chinese, Japanese). With the 'Disaster of 1898,' Spain lost most of its remaining 'colonies after the Empire' (Fradera, 2005). The Chinese Empire was suppressed in 1911, the Russian in 1917, the German and the Austro-Hungarian after the defeat in WWI. In 1936, there were only three formal emperors in the world: the long-standing Japanese and British monarchs, together with the newly declared Italian Emperor after the Italo-Ethiopian War. Nonetheless, the French Republic, the Spanish nationalist government, the United States, the monarch of Belgium (as personal owner) and the Kingdom of the Netherlands maintained substantial colonies and dominions. Italy lost its empire during WWII, although the protectorate over Somalia survived until 1960, the same year in which Congo became independent from Belgium. The French Republic faced two wars before renouncing Indochina and Algeria. Southern European dictatorships maintained overseas colonies: Equatorial Guinea was a colony of Francoist Spain till 1968, and Angola became independent in 1975, after the April Captains overthrew Salazar's dictatorship.

Most historians working on science and medicine in relation with colonialism have taken the British rule of South Asia as a model. Ever since Karl Marx's analysis of British capitalism (Marx, 1867–1894), the centrality of India for the British World System has been acknowledged. Furthermore, historians of tropical medicine are often influenced by postcolonial theorists, who mostly focus on South Asia and its relationship with British colonialism (Chakravorty Spivak, 1999; Chakrabarti, 2000; Sartori, 2008). Mark Harrison has argued that the experience in India led the British to grow their fears for racial degeneration and to abandon the idea that the Europeans could adapt to Indian climate (Harrison, 1999). Michael A. Osborne has argued that this model cannot be extended to other European experiences and that, in particular, it does not work for the French case (Osborne, 2014). India was, indeed, in many respects unique. The Mughal Empire already had a modern state structure, unlikely other colonial dominions. Furthermore, the British Indian civil service provided the Indian republic with an essential backbone after the demise of the Empire. The Belgian and French colonial administrations also regarded the British Raj as a model. This is why, in order to fully understand how colonial/tropical medicine developed in the context of colonial and imperial nations, we need to take a closer look to the diverse nature of colonialism. Jürgen Osterhammel attempted a classification of colonialism, creating six categories: (1) total migration of entire populations and societies; (2) mass individual migration; (3) border colonization; (4) overseas settlement colonization; (5) empire-building war of conquest and (6) construction of naval networks (Osterhammel, 1995, pp. 4–10; see also Finley 1976).

The Habsburg colonization of Bukovina and Moldavia under Joseph, the Russian expansion towards Central Asia and the New England colonization were all instances of frontier colonization. Colonial power relations could even take place without colonies, such as in the case of England and the Celtic fringe (Wales, Scotland, Ireland), and to a lesser extent Bourbon Spain and Catalonia (Taylor, 1948; Judson, 2016; Osterhammel, 2010; Elliott, 2018). As a last example, migrations from European countries such as Germany and Italy to South America can be considered instances of settlement colonialism, and they were even referred to as 'colonies': the German Liebig colony founded in Argentina in 1924 and supported by the eponymous firm is such a case.

The different shades of imperialism also call for more nuanced analysis of the relationship between national identity and empire-building (Conrad, 2010; Fradera, 2018). In the case of the Habsburg Empire, nationalism and empire conflicted. The same happened in Britain, but for different reasons: according to David Edgerton, the British nation was created with the decline of the Empire since 1945 (Edgerton, 2019). On the other hand, the United States was born from an anti-colonial war, but this did not prevent it to develop an informal commercial and military empire in the Caribbean and the Pacific. However, it took a long time for some of these outposts to be recognized as a part of the nation (Go, 2005). France, despite the universal principles of citizenship affirmed by the Revolution, also created an empire with a hierarchy of citizens. Furthermore, the power relationships within empire were continuously shifting and a simple 'centre-periphery' framework is not enough to describe the network of exchanges within and between empires, nor is a simple domination narrative.

According to A. J. P. Taylor (1948) the Habsburg Monarchy was 'a vast collection of Irelands,' but the relations between these Irelands were not linear. On the eve of its dissolution, for instance, Croatia was subjected to Hungary, rather than to Vienna. In his masterpiece *Der Mann ohne*

Eigenschaften, Robert Musil ironically described the different and even incompatible coexisting notions of state in the Habsburg Monarchy on the eve of the First World War:

> This sense of the Austro-Hungarian state was so oddly put together that it must seem almost hopeless to explain it to anyone who has not experienced it himself. It did not consist of an Austrian part and a Hungarian part that, as one might expect, complemented each other, but of a whole and a part; that is, of a Hungarian and an Austro-Hungarian sense of statehood, the latter to be found in Austria, which in a sense left the Austrian sense of statehood with no country of its own. The Austrian existed only in Hungary, and there as an object of dislike; at home he called himself a national of the kingdoms and lands of the Austro-Hungarian Monarchy as represented in the Imperial Council, meaning that he was an Austrian plus a Hungarian minus that Hungarian, and he did it with enthusiasm but only for the sake of a concept that was repugnant to him, because he could bear the Hungarians as little as they could bear him, which added still another complication to the whole combination. This led many people to simply call themselves Czechs, Poles, Slovenes or Germans, and this was the beginning of that further decay and those well-known 'unpleasant phenomena of an internal political kind,' as Count Leinsdorf called them [...]. (Musil, 1930–43, pp. 180–181)

Furthermore, empires were complex and articulated spaces, where different nationalities often coexisted. The transition from the Ottoman Empire to the modern secular state, for instance, was punctuated by the persecution of ethnic and religious minorities and witnessed one of the worst crimes against humanity of the twentieth century: the genocide of the Armenians (Mann, 2005). The Habsburg monarchy, where Metternich had ruled by means of repression, was still described by Robert Musil as a paternalistic autocratic state. Nonetheless, the Imperial government was determined to respect the cultural rights of all nationalities. It comes, therefore, as no surprise that acute observers such as Joseph Roth and Stefan Zweig fondly remembered the Austro-Hungarian Empire, in which different nationalities lived peacefully.

Financial and commercial interactions between empires must be taken into account too, since economics and politics are obviously mingled. The old and backward Habsburg and Ottoman empires were instrumental to geopolitical stability in Europe in the nineteenth century, so that financial support was always available (Taylor, 1948). British Empire attracted

German capital, and Russian frontier colonialism was financially supported by French capital (Osterhammel, 1995, p. 26; Etkind, 2011).

Modern empires also shaped views about the nation in post-colonial independent African and Asian countries. European authoritarian colonial domination contributed to the creation of post-colonial authoritarian governments in Africa and Asia. The independence leaders inherited the colonial territory-state and converted it into a nation-state, even where a pre-colonial sense of nation was lacking.

After WWII, the decolonization process began. Truman administration policymakers believed that in a world centred on bilateral agreements and closed economic systems totalitarianism flourished. They were persuaded that in the interwar period it was the closing of free international trade that led to the rise of fascism in Europe. The economy of Nazi Europe was organized around Germany by means of a system of bilateral agreements between Germany and all other European countries, whose economies worked to sustain the German economy. By the end of WWII, the Soviet Union was establishing bilateral agreements with Eastern European countries to create an economic space to support its economic and social reconstruction. Thus, once a reconstruction of the European economy pivoting on the French economy had failed, Washington decided to leave aside the Morgenthau plan and set up the European Recovery Program, the so-called Marshall Plan, whose aim was the reconstruction of Western Europe and the harmonization of European economies with that of the United States. The instrument was economic, the goal was political (Milward, 1986; Hogan, 1989; Leffler, 1993). Only in a peaceful and prosperous world would the European working classes and the masses of the developing countries resist the appeal of the liberation promises of communism. Thus, Washington assisted the reconstruction of the Japanese economy and pushed Britain and France to dismantle their colonial empires. By the end of the 1960s, only a handful of formal colonies survived, and informal domination replaced formal empires. Large multinational corporations played a part in the asymmetric relations with former formal and informal colonies in Africa, Asia and South America. At the same time, the United Nations Organization and its affiliate institutions, such as the UNESCO, the FAO and the World Health Organization, were created, providing a crucial space for the Cold War confrontation that was the context for the process of decolonization. During the 1950s and 1960s, in South America large companies such as the United Fruit Company and Standard Oil dismissed their medical health programmes, which were replaced by the

UNICEF, the WHO, and in South America by the PASB (Pan-American Sanitary Bureau) for the improvement of health conditions in developing countries. Likewise, the Rockefeller Foundation shut down its International Health Division in 1951 (Farley, 2004; Cueto, 2008).

Tropical Medicine, Medicine in the Tropics, Medicine, and Colonialism

Tropical/colonial medicine developed within the above-outlined context. In each local context, a narrative of what is an empire and what is a colony was created: although the narratives may have had common features, it is worth underlining their diversity. As a consequence, it is possible to observe colonial and tropical medicine in the various contexts and evaluate how global and local interactions shaped medical concepts and practices.

One approach to tropical medicine is strictly geographic, encompassing all medicine practiced in the tropical areas. But it could also be regarded as the branch of medicine dealing with diseases specific to the tropics, such as sleeping sickness, or yellow fever. Neither definition is, however, fully satisfying. Sir Patrick Manson in his 1898 classic textbook identified tropical diseases by contrasting them with 'cosmopolitan' bacterial infections (cholera, plague, tuberculosis). Tropical diseases were characterized by the complexity of their infection cycle (involving a number of hosts and vectors) and their pathogens, usually protozoans and worms. The ecology of the vector (e.g., in the case of malaria, the *Anopheles* mosquito's habitat) was thus of paramount importance for the categorization of a specific disease. However, other investigators often tended to have a broader view of the topic. Giuseppe Sanarelli's *Manuale di igiene generale e coloniale* (A Textbook of General and Colonial Hygiene), published in 1914, three years after the Italian occupation of Tripolitania and Cyrenaica and on the eve of WWI, devoted chapters to the hygienic measures, food and beverages, climate and tropical diseases that could affect both South America and East Africa (Sanarelli, 1914). From a geographical point of view, Sanarelli's approach focused on diseases and health in tropical and subtropical areas. Nonetheless, his approach was rather a consequence of the connection he made between the colonial expansion in Africa and the massive Italian emigration towards the Americas.

Certain infectious diseases such as malaria, cholera and trachoma were not limited to tropical areas. Malaria was, indeed, endemic in certain areas of Southern Europe, South America, East and Central Asia and sub-tropical regions of the United States and Russia; it also affected temperate regions of Central and Northern Europe and of North America, and it was a significant health and economic factor in infested areas. At the beginning of the nineteenth century, nearly half of the world population was at risk for the disease, but by the end of the century, it was mainly confined to tropical and sub-tropical regions (including Southern Europe); as stressed by Randall Packard (2007), it became in a way a tropical disease, a disease of tropical areas. Leprosy and trachoma are now officially considered by the WHO 'neglected tropical diseases' (https://www.who.int/neglected_diseases/diseases/en/), and they have disappeared from most of Europe. However, in the nineteenth century, leprosy still affected certain areas of Northern Europe such as Norway and Scotland (Gussow & Tracy, 1970; Edmond, 2006), and trachoma was probably the main cause of rejection of Italian immigrants to the United States. Typhus and cholera epidemics periodically erupted in Europe, above all in Central and Eastern Europe. As Paul Weindling has reconstructed, the fight against typhus and cholera was strictly connected to German colonialism in the East (Weindling, 2000 p. xix).

The interaction between global politics—war, colonialism, trade—and biomedical research took several forms. Quinine—isolated from cinchona bark in 1820—has been considered a 'tool of empire' for the European penetration of Africa (Headrick, 1981), although that interpretation has been questioned (Chakrabarti, 2014). Typical colonial problems stimulated research. The Japanese occupation of Java (the Dutch colony hosting most of the cinchona plantations) in 1942 meant a shortage of quinine, at a time when US troops were fighting the war largely in malarial areas: North Africa, Southern Italy, the Pacific theatre. This was a powerful factor in setting up a research programme on synthetic anti-malarials—ultimately resulting in the production of chloroquine (Slater, 2009). Vannevar Bush's *Science: The Endless Frontier* (Bush, 1945) mentioned chloroquine together with the large-scale production of penicillin, achieved in 1943. Both were the result of the war effort, and both were of help in establishing an informal US empire after WWII.

Colonial relationships could also be an exchange, with practices and concepts travelling both ways. This happened, for example, in the United States according to Warwick Anderson (2006), with the diaspora of American colonial physicians from the Philippines, and the great reputation gained by the heroes of colonial medicine such as William H. Gorgas.

From the colonies, the same civilizing mission extended to minorities or the underclass in the homeland. But while in the past these social categories were directly accused of spreading diseases, now they became a disease to be controlled whatever it took. The pursuit of health became crucial in the construction of an 'imagined community': home and abroad, the metaphor of the 'political body' of the nation circulated widely (Martin, 1990; Weindling, 2000; Craddock & Dorn, 2001; Latour, 2001; Bashford, 2004).

After WWII, the spaces of international relations radically changed. The newly created international agencies of the United Nations such as the UNESCO, the FAO, the UNICEF and the WHO became central. The WHO replaced the Rockefeller Foundation in the early post-WWII period in planning and managing specific programmes targeting global health objectives, such as the eradication of malaria. The top-down, colonial and technocratic approach was largely directed by the US interests. In the 60s, the Soviet Union's suspicious attitude changed, so that the WHO started to play a part in the multilateral negotiation spaces of international cooperation, and public health in Third World countries became another Cold War territory, also as a result of the demise of the European empires. The 1978 Declaration of Alma-Ata probably marked the end of the old public health world order, with a shift towards the implementation of primary health services with more autonomy granted to local institutions and governments. The United States also used national agencies such as the ICA of the Department of State and later on USAID to foster health in developing countries (Cueto, 2008; Cueto et al., 2019).

Medicine, Empire, the Nation

Migration medicine is another perspective to be used to look at the special relation between nation and biomedical practices. Scientists have played a part in the processes of people in motion since the last three decades of the nineteenth century. Controlling migration flows involved the ability to identify individuals with a certain degree of certitude, and carefully vetting individuals and groups for potential sanitary threats. As noted above, Bertillonism catered for the first aim. Under the Pasteurian approach, health concerns were addressed by deploying any sort of weapons against the spread of germs (Conrad, 2010; Weindling, 2000). However, Pasteur was to be respected as long as international trade was not harmed. Britain approached public health mostly by promoting sanitation: quarantine and isolation were considered a threat to international trade, as well as to the concentration of workers in industrial towns (Baldwin, 1999, pp. 28–29).

As of the mid-1800s, novel international spaces opened up for the international management of health crises: in 1851, the first International Sanitary Conference was held in Paris in order to address the threat posed by cholera, plague and yellow fever (see Howard-Jones, 1975; Baldwin, 1999, Rasmussen, 2001). The increasing globalization of public health also called for a global elite of mostly liberal physicians and medical investigators sharing the civilizing mission of Western medicine, meeting at conferences, conferring in local health boards of international relevance or competing in the field. For example, the conflict between Pasteur's and Koch's schools over the 1883 cholera epidemics in Egypt was part of the larger picture: the French saw the chance to gain a foothold in the area, where they had scarce influence after the establishment of the British rule the year before. Koch isolated and identified the cholera *Vibrio* in India, and Émile Roux admitted the defeat. Upon his return to Berlin, Koch was acclaimed as a hero: German supremacy had been established once again after the victory at Sedan (Brock, 1988; Gradmann, 2009). A decade later, the British abided by the Venice International Convention, and sanitary control over Egypt—the most important gateway to the Mediterranean—was eventually shared among several European countries, each of them with its own agenda going beyond the preservation of public health. Throughout the long nineteenth century, the Europeans took divergent paths as regards the fight against infective diseases. Different attitudes were developed over time, and the same government or health office might move from one strategy to another, according to convenience, the target disease and shifting medical theories. Nonetheless, some patterns can be drawn. The British followed radical sanitationism and the Prussians tempered quarantinism. Strict control over prostitution was adopted on the Continent, but not in Britain or in Argentina. Compulsory smallpox vaccination was implemented in France, Germany and Brazil (Ackerknecht, 1948; Kuhnke, 1990; Worboys, 2001; Baldwin, 1999). Likewise, a number of international meeting and regulation spaces were created in which the European powers and the United States clashed. In 1902, the United States promoted the creation of the *Oficina Panamericana de Sanidad*, whose aim was to prevent the spreading of diseases from Europe to South America, but which also aimed to contain European influence in the American continent. Five years later, in 1907, the *Office international d'hygiène publique* (OIHP) was created to ratify sanitary agreements among states (Cueto, 2004).

The elite of physicians and biomedical scientists participated in the justification of European domination as a consequence of the moral supremacy of the white man. Western medicine at the end of the nineteenth century could finally provide some successes, thanks to germ theory and new effective drugs. The new medicine was a part of the civilizing mission and was often used to support free trade and peaceful colonial penetration.

Following the sixteenth and seventeenth-century British and Spanish theorists, nineteenth-century formal and informal empires found their justification in some sort of moral superiority of the white man, whose mission consisted in bringing civilization to inferior societies. Indeed, doctors usually belonged to liberal elites, being cosmopolitan and pacifist, but mostly in favour of the civilizing missions. In Britain, the liberals were for free trade and non-military economic expansion (Padgen, 1998; Darwin, 2009; Edgerton, 2019). When dealing with 'the other,' physicians showed a wide variety of approaches, depending on their views and their social and cultural position. Doctors often presented themselves as agents of peaceful colonial penetration. In 1897, Dr. Accurso, a physician of the Italian Royal Navy, reporting on Benadir, observed that the natives mistrusted missionaries and explorers, but often attended Western doctors. Dr. Bernhard Dernburg in his *Zielpunkte der Deutschen Kolonialwesen* explicitly theorized the function of Western doctors in peaceful colonial penetration. By contrast, during the cholera epidemic of 1902 in the Philippines, US army surgeons were even stricter than military officials in implementing health measures they deemed urgent and were disrespectful of local knowledge and experience. When American civil health service officials replaced military doctors, the health inspector presented the colonial domination relation as a softer civilizing mission, paving the way for the indirect dominion relation of postwar decolonization (Accurso, 1897; Ileto, 1988; Sullivan & Ileto, 1997; Anderson, 2006; Conrad, 2008, p. 64).

The role of physicians was thus multifaceted. Poets, writers, historians and intellectuals were those who created national identities (Taylor, 1948; Anderson, 1983; Hobsbawm, 1990). Physicians were sometimes poets, writers, intellectuals and politicians. In the nineteenth century, nationalism spread from towns to the countryside, as peasants learned to read. Newspapers and schools were crucial instruments for the making of the 'imagined community.' Global interactions and the formation of the idea of the nation were not in opposition (Conrad, 2010). Doctors in a way often participated in this process by spreading the idea of a civilizing mission of their motherland combined with a universal mission of science and medicine. In South America, it was the elites of the coastal metropolises who spread the idea of a mission to civilize the hinterland populations. Civilizing discourses were not only an instrument of domination of the political and hegemonic groups/regions but also a part of the process of nation-building.

The case of Louis Pasteur is paramount. He and his entourage actively constructed his twofold image as a universal hero of mankind and a nationalistic character in the global geopolitical confrontation with Germany.

Medicine, Empire, and the Nation: The Essays

Following the outlined view of the complex relationships between the nation, empire, colonies and medicine, we propose to revise the approach to tropical medicine and include any research and practice connecting the study of medicine to colonial expansion and the dominance relations of the decolonization world post-WWII. The essays in this volume touch upon a number of facets of the complex relationship between tropical medicine, now broadly construed, and the making of the empire and the nation, between the Scramble for Africa and the 1960s decolonization, as outlined in this introduction. Taken together, they aim to contribute to shed light on the role played by tropical medicine in the complex interaction between nation and empire. Though all the essays revolve around a common central issue, we can divide them into three groups. A first group deals with the role of tropical medicine in the evolution and collapse of empires. Isabel Amaral identifies tropical medicine as a factor of the consolidation of Portuguese imperial power in the African colonies, addressing the public health programmes directed at the control and the eradication of sleeping sickness from the early decades of the twentieth century to the post-WWII period. However, health programmes after WWII had also unexpected consequences, as they ended up promoting anti-colonial movements and the strive for self-determination.

Stephen Snelders focuses on the fight against leprosy and drug abuse in the Dutch West Indies as part of a colonial policy, aimed at protecting the imperial project. Drug abuse and leprosy are usually considered in the metropolis two quite different issues both in terms of public health and public order. Nevertheless, their framing and management showed striking similarities in the colonies. Indeed, in both cases, non-white leprosy sufferers and drug users from different non-Western backgrounds were constructed by colonial medicine as passive and undisciplined. They were seen as a nuisance for labour efficiency and a threat to Europeans, contaminating the Western race (by means of sex or disease) and inducing the white men to 'go native.'

A second group of essays analyses the relations between a number of tropical medical institutions and imperial commercial and political expansion. The relationship between the international business community and medical investigators in tropical medicine is the focus of Deborah J. Neill's and Juanita De Barros' essay. Indeed, they argue that support for transnational corporate interests was the most fundamental factor in the establishment of the Liverpool School of Tropical Medicine. Trading, plantation and shipping companies from Britain, France, Belgium and elsewhere funded and supported the School in the hope of solving the problems of tropical diseases and thereby expanding their economic reach in the Global South. They show that, although there were some common ideas beyond the measures undertaken to fight malaria, local politics and local elites also played an important part. Ideas and policies locally implemented circulated within the imperial space. Jaime Larry Benchimol focuses on the work of Brazilian researchers on leishmaniasis, a major health problem in Brazil since the 1930s, leading to the consolidation of the concept of American tegumentary leishmaniasis. The chapter examines the knowledge production and the efforts to control this disease that have mobilized health professionals, scientists, government agencies and institutions, non-governmental organizations, international agencies and rural and urban populations at regional, national and global levels. This chapter frames the Brazilian case within the context of the decolonization process, as leishmaniasis apparently spread in a number of countries in Africa, Asia and South America in the 1950s and 1960s when they were gaining independence.

Fifteen years ago, a seminal paper by Jim Secord, 'Knowledge in Transit,' brought to the attention of historians of science that knowledge is intrinsically defined by its being communicated, circulated, adapted and elaborated (Secord, 2004). Circulation of people became a crucial factor following the resumption of European colonial expansion. In the introduction to the *Age of Empire*, Eric Hobsbawm recalled how the fact that he was born in Alexandria in 1917 to a British official of Russian Polish origin who worked in the Egyptian Post and Telegraph Service and a

Viennese young lady on a leisure trip after graduation was something that can be easily explained in the light of the history of modern empires (Hobsbawm, 1987). Thus, a third group of essays tackles the crucial issue of the inter-related circulation of people, objects and ideas as something constitutive of the history of modern empires. Mauro Capocci and Daniele Cozzoli explore the relations between Italian colonial expansion and emigration through the reports of Navy doctors, who worked aboard both ships sent on military missions around the globe and on emigration service vessels between 1885 and 1915. The analysis of the work of Navy physicians allows the historian to see the complex relationships between colonialism, emigration and the creation of the Italian nation in the context of the development of the European colonial empires. The authors stress how the chosen perspective highlights how Italian immigrants in South America can play both the role of subalterns regards the metropolitan elites and colonizers regards the local communities. Matheus Duarte da Silva's chapter deals with the knowledge of plague's environment. Focussing on Brazil, the chapter details how medical researchers had to enlarge their view on the disease to encompass a 'sylvatic' plague beside the more common 'urban' infection. The understating of plague as a geographically stable disease—endemic in specific animal populations—as opposed to a mobile threat mostly spread by rats aboard the ships was connected in Brazil to the construction of the central nation-state in the twentieth century and to the transformation of plague in a tropical disease.

Carles Brasó's chapter sets forth a parallel between war and medicine, with multiple cross-references among air strikes in the 30–40s and the medical knowledge circulating in settings as diverse as Spain, India and China. The campaign for banning air strikes mobilized several actors within the anti-fascist, and later anti-colonial, field. In this global network, medical knowledge travelled and was appropriated in different contexts, with concrete innovations such as the creation of blood banks and the deployment of mobile surgical units at the front.

References

About, I. (2011). Surveillance des identités et régime colonial en Indochine, 1890–1912. *Criminocorpus* [Online], Identification, contrôle et surveillance des personnes, Articles, Online since 23 May 2011, connection on 06 November 2019. http://journals.openedition.org/criminocorpus/417

Accurso, S. (1897). Brevi cenni sulle condizioni climatico-igieniche del Benadir. *Annali di Medicina Navale*.
Ackerknecht, E. H. (1948). Anticontagionism Between 1821 and 1867. *Bullettin of the History of Medicine, 22*(5), 562–593.
Anderson, B. (1983). *Imagined Communities*. Verso.
Anderson, W. (2006). *Colonial Pathologies. American Tropical Medicine, Race and Hygiene in the Philippines*. Duke University Press.
Baldwin, P. (1999). *Contagion and the State in Europe 1830–1930*. Cambridge University Press.
Bashford, A. (2004). *Imperial Hygiene: A Critical History of Colonialism, Nationalism and Public Health*. Palgrave Macmillan.
Bayly, C. (2004). *The Birth of the Modern World*. Blackwell.
Benchimol, J. L. (1999). *Dos micróbos aos mosquitos. Fevre amarela e a revoluçao pasteuriana no Brasil*. Editora Fiocruz.
Bhattacharya, S., Harrison, M., & Worboys, M. (2005). *Fractured States: Smallpox, Public Health, and Vaccination Policy in British India, 1800–1947*. Orient Blackswan.
Bovet, D. (1989). *Une chimie qui guérit*. Payot.
Brock, T. (1988). *Robert Koch. A Life in Medicine and Bacteriology*. University of Wisconsin Press.
Bush, V. (1945). *Science, The Endless Frontier. A Report to the President*. United States Government Printing Office.
Chakravorty Spivak, G. (1999). *A Critique of Postcolonial Reason. Towards a History of the Vanishing Present*. Harvard University Press.
Chakrabarti, D. (2000). *Provincializing Europe. Postcolonial Thought and Historical Difference*. Princeton University Press.
Chakrabarti, P. (2014). *Medicine and empire, 1600–1960*. Palgrave MacMillan.
Conrad, S. (2008). *Deutsche Kolonialgeschichte*. Beck (trans. into English as *German Colonialism*. Cambridge University Press, 2012.
Conrad, S. (2010). *Globalization and the Nation in Imperial Germany*. Cambridge University Press.
Cozzoli, D. (2018). The Post-colonial Context of Daniel Bovet's Research on Curare. In M. Savelli & P. Manning (Eds.), *Global Transformations in the Life Sciences, 1945–1980* (pp. 31–46). Pittsburgh University Press.
Craddock, S., & Dorn, M. (2001). Nationbuilding: Gender, Race and Medical Discourse. *Journal of Historical Geography, 27*, 313–318.
Crozier, A. (2007). *Practicing Colonial Medicine: The Colonial Medical Service in British East Africa*. Bloonsbury.
Cueto, M. (2004). *El valor de la salud: historia de la Organización Panamericana de la Salud*. Organización panamericana de la Salud.
Cueto, M. (2008). International Health, the Early Cold War and Latin America. *CBMH/BCHM, 25*(1), 17–41.

Cueto, M., Brown, T. M., & Fee, E. (2019). *The World Health Organization: A History*. Cambridge University Press.

Darwin, J. (2009). *The Empire Project: The Rise and Fall of the British World System 1830–1970*. Cambridge University Press.

Dedet, J.-P. (2000). *Les Instituts Pasteurs d'outremer*. l'Harmattan.

Delaunay, R. (1962). *L'Institut Pasteur des origines a aujourd'hui*. Éditions France-Empire.

Di Nolfo, E. (2007). *Dagli imperi militari agli imperi tecnologici. La politica internazionale dal XX secolo ad oggi*. Laterza.

Di Nolfo, E. (2015). *Storia delle Relazioni Internazionali*. Laterza.

Dubos, R. (1995). *Louis Pasteur Franc-tireur de la science*. La Découverte.

Edgerton, D. (2019). *The Rise and Fall of the British Nation*. Penguin.

Edmond, R. (2006). *Leprosy and Empire. A Medical and Cultural History*. Cambridge University Press.

Eichengreen, B. (1992). *Golden Fetters. The Gold Standard and the Great Depression 1919–39*. Oxford University Press.

Elliott, J. H. (2018). *Scots and Catalan. Union and Disunion*. Yale University Press.

Etkind, A. (2011). *Internal Colonization. Russia's Imperial*. Polity Press.

Farley, J. (2004). *To Cast Out Disease: A History of the International Health Division of the Rockefeller Foundation, 1913–1951*. Oxford University Press.

Ferguson, N. (2004). *Empire. The Rise and Demise of the British World Order and the Lesson for Global Power*. Basic Books.

Finley, M. (1976). Colonies: An Attempt at a Typology. *Transactions of the Royal Historical Society, 26*, 67–188.

Fradera, J.-M. (2005). *Colonias para después de un Imperio*. Bellaterra.

Fradera, J.-M. (2018). *The Imperial Nation*. Princeton University Press.

Friedman, N., & Schwartz, A. J. (1963). *A Monetary History of the United States*. Princeton University Press.

Galeano, D., & García Ferrari, M. (2011). The *bertillonage* in the South American Atlantic World. *Criminocorpus* [Online], Identification, contrôle et surveillance des personnes, Articles, Online Since 19 May 2011, Connection on 06 November 2019. http://journals.openedition.org/criminocorpus/402

Geison, G. (1995). *The Private Science of Louis Pasteur*. Princeton University Press.

Go, J. (2005). Imperial Power and Its Limits: America's Colonial Empire in the Early Twentieth Century. In C. J. Calhoun, F. Cooper, & K. W. Moore (Eds.), *Lessons of Empire: Imperial Histories and American Power* (pp. 201–214). The New Press.

Gradmann, C. (2009). *Laboratory Disease. Robert Koch's Medial Bacteriology*. The Johns Hopkins University Press.

Greenwood, A. (Ed.). (2016). *Beyond the State: The Colonial Medical Service in British Africa*. Manchester University Press.

Greenwood, A., & Topiwala, H. (Eds.). (2015). *Indian Doctors in Kenya, 1895–1940*. Palgrave Macmillan.

Gussow, Z., & Tracy, G. S. (1970). Stigma and the Leprosy: The Social History of a Disease in the Nineteenth and Twentieth Century. *The Bulletin of the History of Medicine, 44*, 425–449.

Hamlin, C. (2009). *Cholera: The Biography*. Oxford University Press.

Harrison, M. (1999). *Climates and Constitutions: Health, Race, Environment and British Imperialism in India (1650–1830)*. Oxford University Press.

Haynes, D. M. (2017). *Fit to Practice: Empire, Race, Gender, and the Making of British Medicine, 1850–1980*. University of Rochester Press.

Headrick, D. A. (1981). *The tools of empire: Technology and european imperialism in the nineteenth century*. Oxford University Press.

Hobsbawm, E. (1987). *Age of Empire*. Vintage Books.

Hobsbawm, E. (1990). *Nation and Nationalism*. Cambridge University Press.

Hobsbawm, E. (1996). *Industry and Empire*. Penguin.

Hochstadt, S. (1999). *Mobility and Modernity*. Michigan University Press.

Hogan, M. J. (1989). *The Marshall Plan. America, Britain and the Reconstruction of Western Europe, 1947–1952*. Cambridge University Press.

Howard-Jones, N. (1975). *The Scientific Background of the International Sanitary Conferences, 1851–1938*. World Health Organization.

Ileto, R. C. (1988). Cholera and the Origins of American Sanitary Order in the Philippines. In D. Arnold (Ed.), *Imperial Medicine and Indigenous Societies* (pp. 125–148). Manchester University Press.

Johnson, R. (2010). The West African Medical Staff and the Administration of Imperial Tropical Medicine, 1902–14. *The Journal of Imperial and Commonwealth History, 38*(3), 419–439.

Judson, P. (2016). *The Habsburg Empire. A New History*. Harvard University Press.

Kuhnke, L. (1990). *Lives at Risk: Public Health in Nineteenth-Century Egypt*. University of California Press.

Latour, B. (2001). *Pasteur: Guerre et Paix des microbes*. La Découverte.

Leffler, M. (1993). *A Preponderance of Power. The Truman Administration and the Cold War*. Stanford University Press.

Löwy, I. (1990). Yellow Fever in Rio de Janeiro and the Pasteur Institute Mission: A Case Study of Transfer of Science to the Periphery. *Medical History, 34*, 144–164.

Löwy, I. (2001). *Virus, moustiques et modernité: la fièvre jaune au Brésil entre science et politique*. Éditions des Archives contemporaines.

Mann, M. (2005). *The Dark Side of Democracy: Explaining Ethnic Cleansing*. Cambridge University Press.

Martin, E. (1990). Toward an Anthropology of Immunology: The Body as Nation-State. *Medical Anthropology Quarterly, 4*, 410–426.

Marx, K. (1867–1894). *Das Kapital.* In K. Marx & F. Engels, *Gesamtausgabe* MEGA Berlin: De Gruyter (trans. into English as Karl Marx *The Capital.* Penguin, 1990).
McNeill, J. R. (2010). *Mosquitos Empires. Ecology and War in the Great Caribbean, 1620–1914.* Cambridge University Press.
Milward, A. (1986). *The Reconstruction of Western Europe.* Routledge.
Moulin, A. M. (1992). *Patriarchal Science.* In P. Petitjean, C. Jami, & A.-M. Moulin (Eds.), *Science and Empire. Historical Studies About Scientific Development and European Expansion.* Springer.
Musil, R. (1930–1943). *Der Mann ohne Eigenschaften.* Rowolth (trans. into English as *The Man Without Qualities.* Picador, 1995).
Neill, D. (2012). *Networks in Tropical Medicine: Internationalism, Colonialism, and the Rise of a Medical Specialty, 1890–1930.* Stanford University Press.
Nicolle, C. (1935). *Responsabilité de la Médecine dans l'œuvre d'extension de la civilization.* In Nicolle (Ed.), *Responsabilités de la Médecine* (pp. 37–53). Félix Alcan.
Osborne, M. (2014). *The Emergence of Tropical Medicine in France.* University of Chicago Press.
Osterhammel, J. (1995). *Kolonialismus. Geschichte, Formen, Folgen.* Beck (trans. into English as *Colonialism.* Princeton University Press, 2005).
Osterhammel, J. (2010). *Die Verwandlung der Welt: eine Geschichte des 19. Jahrhunderts.* Beck (trans. into English as *The Transformation of the World: A Global History of the Nineteenth Century.* Princeton University Press, 2015).
Packard, R. (2007). *The Making of a Tropical Disease. A Short History of Malaria.* Johns Hopkins University Press.
Padgen, A. (1998). *Lords of all the world: Ideologies of empire in Spain, Britain and France c.1500–c.1800.* Yale University Press.
Paquette, G. (2020). *The European Seaborne Empires.* Yale University Press.
Pati, B., & Harrison, M. (2009). *The Social History of Health and Medicine in Colonial India.* Routledge.
Pope, J. W. (1924). Synthetic Therapeutic Agents. *British Medical Journal, 1*, 413–414.
Rasmussen, A. (2001). *L'Hygiène en Congrès* in Bourdelais ed. *Les Hygiénistes: Enjeux, modèles et pratiques.* Belin, pp. 213–239.
Sanarelli, G. (1914). *Manuale d'igiene generale e coloniale.* Barbera.
Sartori, A. (2008). *Bengal in Global Concept History: Culturalism in the Age of Capital.* Chicago University Press.
Secord, J. (2004). Knowledge in Transit. *Isis, 95*(4), 654–672.
Slater, L. (2009). *War and Disease.* Rutgers University Press.
Solomon Gross, S. (2006). *Doing Medicine Together: Germany and Russia Between the Wars.* University of Toronto Press.

Sullivan, R. J., & Ileto, R. C. (1997). Americanism and the Politics of Health in the Philippines 1902–13. In S. Heva & P. Hove (Eds.), *Philanthropy and Cultural Context* (pp. 39–64). University Press of America.

Taylor, A. J. P. (1948). *The Hapsburgh Monarchy 1809–1918. A History of the Austrian Empire and Austria-Hungary*. Hamish Hamilton.

Wallerstein, I. (2004). *World-System Analysis: An Introduction*. The John's Hopkins University Press.

Wallerstein, I. (2011). *The Modern World-System 3 vols*. University of California Press.

Weindling, P. (2000). *Epidemics and Genocide in Eastern Europe*. Oxford University Press.

Worboys, M. (2001). The Colonial World as Mission and Mandate: Leprosy and Empire, 1900–1940. *Osiris, 15*, 207–218.

PART I

Tropical Medicine in the Evolution and the Collapse of Empires

CHAPTER 2

Tropical Medicine and the "Consolidation" of the Portuguese Empire (1902–1966)

Isabel Amaral

Tropical Medicine: Fulfilling Two Agendas

The emergence of tropical medicine can be understood from different perspectives and historiographical approaches. In this case, a discussion framework is applied to better understand the emergence of scientific specialties between the end of the nineteenth century and the first decades of the twentieth. The study of the emergence of new disciplines or scientific specialties has been discussed since the 1970s by sociologists and

This work is financed by Portuguese national funds through FCT – Fundação para a Ciência e a Tecnologia, I.P. (PIDDAC/OE), within the CIUHCT Unit, Ref.: UIDB/00286/2020.

I. Amaral (✉)
Interuniversity Center for the History of Science and Technology (CIUHCT), NOVA School of Science and Technology, Almada, Portugal
e-mail: ima@fct.unl.pt

M. Capocci, D. Cozzoli (eds.), *Empire, Nation-building, and the Age of Tropical Medicine, 1885–1960*, Medicine and Biomedical Sciences in Modern History,
https://doi.org/10.1007/978-3-031-38805-7_2

historians (Lemaine et al., 1976; Amaral, 2006). Tropical medicine became a recognized field of teaching, research, and professional practice in the twentieth century, as identified by Michael Worboys (1976). It acquired the status of an independent discipline, with an object of study, a language, and a methodology distinct from the general medicine practiced by Portuguese military physicians in the colonies since the sixteenth century.

In the historiography of the Third Portuguese Colonial Empire,[1] there continue to be gaps in the discussion about the role of tropical medicine in the construction and development of the colonial project in the twentieth century. Some authors have focused on the circulation and appropriation of medical knowledge (Shapiro, 1983; Neill, 2012; Amaral, 2020), others on the impact and reaction of vertical health policies overseas on the native population (Tilley, 2011, Coghe, 2017; Havik et al., 2017; Fanon, 2020), and still others on the boundaries between conventional European medicine and folk medicine. This study aims to examine how, by controlling tropical diseases using tropical medicine and their colonial authority, control of the territory was also achieved.

The aim of this study is to examine how tropical medicine's control of tropical diseases not only established a new area of research with scientific authority but also became a tool of empires, as noted by Daniel Headrick in the introduction (Headrick, 1981). Consequently, tropical medicine also assumed the role of colonial authority by controlling occupied territories, as exemplified in the case of Portugal.

[1] The beginning of the "Third Portuguese Empire"—the shift toward Africa—was a consequence of the independence of Brazil. The Portuguese expeditions between 1840 and 1850 tried to map the African hinterland in such a way as to claim sovereignty over large tracts of territory between Angola and the Indian Ocean coast. Succeeding the conference of Berlin (initiated in 1884) which divided Africa in service to European colonialism, Britain presented Portugal with an ultimatum in 1890, proclaiming that the colonies should be permanently occupied. The end of the Portuguese monarchy and the beginning of the First Republic (which took place between 1910 and 1926) were characterized by an effort to gain control over its African possessions. But it was only with the dictatorial regime, which began in 1926 (and which would last until 1974), that Portuguese colonialism would be established in the modern sense of the term, as an institution, knowledge, and discourse. See Valentim Alexandre and Jill Dias, Nova História da Expansão Portuguesa. O Império africano (A New History of Portuguese Expansion. The African Empire), 1825–1890, Lisbon, Estampa, vol X, 1998 or Alexandre, Valentim, Vellho Brasil, Novas Áfricas—Portugal e o Império (1808–1975) (Lisboa et al., 2000).

The defense of the Portuguese colonial empire in Africa after the British Ultimatum in 1890 was never an easy task for the national political order. Despite efforts made by the different political regimes in Portugal from 1902 to 1966 (monarchy, liberalism, and dictatorship), tropical medicine was always seen as a power variable, inasmuch as it established a privileged dialogue with the local population and local authorities, the power established in Lisbon, and the network of frontier agents in imperial colonies in Africa.

The consolidation of tropical medicine as a field of scientific and medical expertise led to the consolidation of imperial power. This chapter intends to reflect on the role of the Portuguese school of tropical medicine and its actors (humans and non-humans) in the European context and attempts to identify the unconventional cycles of knowledge production and power which led to the longevity of the Portuguese Empire (Oliveira e Costa, 2014).

The Foundation of the Lisbon School of Tropical Medicine Under Royal Rule

Portugal's scientific alignment with other European countries can be traced back to three historical moments in Portuguese medical history: first, the reform of the Naval School course programs and the training of naval doctors in hygiene disciplines and exotic pathology; second, the foundation of the bacteriology office at the Naval School; and, finally, the first mission to Angola to study sleeping sickness in 1901.

The Naval School was responsible for the specialized training of doctors who had been chosen for colonization missions in the African territories between the end of the nineteenth century and the beginning of the twentieth. With tropical medicine still trying to establish its own cognitive space between miasma theory, Pasteurian bacteriology, and Patrick Manson's parasitological theory, research into the etiology of diseases assumed a prominent role in the European agenda, starting with sleeping sickness, whose prevalence seriously compromised the colonial project.

The Africans, whose ancestors had lived on the continent for hundreds of thousands (if not millions) of years, knew the disease very well, which also affected cattle. They also associated it with the fly known as "canoe fly" because it was found near rivers, or "elephant fly", and they were also able to detect the disease in humans by the enlargement of the cervical ganglion (identified by Thomas Winterbottom in the eighteenth century), which was of key importance for slave traders to be able to identify sufferers (Steverding, 2008).

Whereas Africa was considered the graveyard of Europeans (Curtin, 1998) because they were particularly affected by malaria, sleeping sickness would have become the most dreaded disease for Europeans because it severely depleted the African workforce. Sleeping sickness was one of the few African diseases affecting African people and was consequently a "truly" tropical disease (Lyons, 1992).

The forced migrations imposed by European empires onto African populations, dictated by the construction of infrastructure such as railways, caused these diseases to become epidemic due to the displacement of patients, pathogenic agents, and insect vectors. As with other colonial powers—the United Kingdom, France, or Belgium—sleeping sickness gradually became the main focus of Portuguese medical and political attention.

As a result of the colonialist agenda, the disease thus assumed an important role within the international medical community. It was, therefore, one of the most studied tropical diseases in Europe (Sabbe-Clarke et al., 1980). Given the importance of this typically African disease, several European medical missions involving specialists of different nationalities were sent to African colonies. The difficulty in determining the etiological agent gave rise to a scientific controversy between the Portuguese and British missions (1898 and 1904) over the identification of Trypanosoma as the cause of the disease (Amaral, 2012).

In 1900, Otto Gleim, a German consul in Luanda, reported the alarming evolution of sleeping sickness in Angola as a result of the construction of the railway from Luanda to Ambaca. At the same time, he denounced the lack of commitment from the Portuguese health authorities in the territory. The Lisbon Society of Medical Sciences, upon becoming aware of this report, immediately responded and convinced the Minister of Maritime and Foreign Affairs that it was necessary to send a medical mission to the colonies to study the disease in situ, in accordance with the principles declared at the Berlin Conference.

The Lisbon Society of Medical Sciences took advantage of the emergency in order to change the colonial agenda, in which medicine would occupy a central place and have a decisive influence on the Portuguese state. To highlight the relevance of Portuguese medicine's participation in the colonial project, the Society went further by emphasizing the importance of medical missions as a guarantor of European colonization in Africa (Doença do Somno, 1900, p. 268; Amaral, 2012):

> Germany, the United Kingdom, France, and Austria have increased the number of their scientific expeditions, and Africa and Asia have seen learned men from European laboratories tear themselves away from the comforts of life to go to these far-off and sometimes barbarous locations, and in the face of dangers and inclement climates to study at close quarters some of these terrible illnesses, which are the scourge of humanity.

Within the framework of imperialist rhetoric, doctors assumed a position of scientific supremacy and moral authority, insofar as they would be willing to give up their home comforts and acquire the necessary strategies to overcome the sanitary difficulties posed by the inhospitable environment, which was understood to be the main obstacle to effective colonization.

In 1901, the Ministry of Maritime and Foreign Affairs sent the first mission to study sleeping sickness in Angola, the jewel in the crown of the empire, and it was in charge of analyzing the disease in this territory. The Royal Society of London supported the first British missions to study sleeping sickness in Entebbe (the first in 1902 and the second in 1903).

This mission supported by the Portuguese state made use of resources which would have been unthinkable in other circumstances, given the economic constraints that the country was struggling with at the time (Amaral, 2012). In fact, this mission had obvious advantages for both parties involved: for the State, which, in a context of international competition for African territories, demonstrated its commitment to controlling the advance of the disease, which directly proved its effectiveness in occupying an African territory; and for doctors, who could explore the wider opportunities for professional recognition which were opened up.

The most renowned researchers in the field of microbiology and tropical medicine were recruited: Annibal Bettencourt, director of the Câmara Pestana Royal Institute of Bacteriology, CPRIB, and head of the mission; Anníbal Celestino Correia Mendes, director of the bacteriological laboratory at the Hospital de Luanda; Ayres Kopke, naval physician, director of the microbiological laboratory at the Naval School; and José Gomes de Rezende Junior, military physician, and João Brás de Gouveia, a researcher at the CPRIB. It was the first European mission to study the disease on African territory and became an unrivaled example of a living laboratory as a result.

The first results of the mission, which indicated that the causal agent of the disease was a bacterium (hypnococcus), were initially published in the first report sent to the Ministry of Maritime and Foreign Affairs in 1901 (Relatórios, 1901). Immediately after the publication of these results, Aldo Castellani, who was working on behalf of the British mission, credited himself as having discovered that the bacterium was the etiological agent of the disease. Therefore, a controversy was established between the Portuguese and British missions (Amaral, 2012).

Considered national heroes, the Portuguese doctors were decorated by King D. Carlos I. The encouraging conclusions of the Portuguese mission were presented at a public conference at the Lisbon Geography Society. The Minister of Maritime and Foreign Affairs praised the Society of Medical Sciences for the mission, thanking them for the influence they had had on the State, and celebrating the glory of Portugal on the European medical scene. The reports produced by the scientific mission were testimony of "a colossal triumph for Portuguese medicine" (Amaral, 2012).

In 1904, the controversy was explained by Castellani. He identified trypanosome as the causal agent of the disease, and thereby distanced himself from the first reaction to the results of the Portuguese team, but in the process initiated a rivalry with his teammates in the United Kingdom. The above events and the scientific controversy, significant as they were, illustrate the ensuing tensions and maneuvering that could arise around a disease. From then on, international attention turned to the establishment of strategic alliances in which Portugal was compelled to participate, both for political and medical and health reasons (Amaral, 2012).

The interests of imperial powers and imperialist ideals were embodied in the opening of the first schools of tropical medicine in the United Kingdom: the Liverpool School of Tropical Medicine in 1898, and the London School of Hygiene and Tropical Medicine in 1899. For the same reason, the medical class, an enlightened and politically influential elite in Portugal, mobilized to acquire scientific and technical knowledge at a specialized training center (a school of tropical medicine), similar to what was happening throughout Europe. This training center *par excellence* would reduce and extinguish the main obstacle to colonization: the diseases that colonizers were unaware of in the temperate zones and whose casualties were greater than those of military conflict (Curtin, 1998).

The foundation of schools of tropical medicine in the United Kingdom could not leave the Portuguese government indifferent. It required the government to take advantage of its extensive overseas territories for the benefit of the new civilization, and create a favorable environment for the foundation of a School of Tropical Medicine, as described by Miguel Bombarda, one of the most influential Portuguese doctors and politicians in parliament in 1901:[2]

> Colonization is not only a social and economic question but also a question of hygiene and pathology. The wealth and prosperity of a colony depend firstly on all the living facilities that may be encountered by the colonizing party. The remedy for the serious risks incurred by colonization undertaken blindly lies in the intervention of medicine, with the highly powerful resources of which it currently disposes of. The United Kingdom, Germany, and France have recently recognized this fact by the creation of study and teaching centers which will simply translate into colonizing facilities and colonial prosperity.

The partition of Africa by European powers was characterized by the impetus of the colonial project. In favor of the effective occupation of Portuguese colonial territory, the myth of "historical rights" prevailed and the colonial project became the central element of Portuguese nationalism. The colonial project, the scientific "occupation" of the territory, and the consolidation of tropical medicine progressed hand in hand.

Defending the primacy of effective colonization in the territory, Portugal committed to aligning itself with European colonialist ideals, trying to defend its possessions, rooted in medical knowledge specializing in exotic pathologies, as justified in the Portuguese parliament in 1902:

> If Portugal must live through and for its colonies, not only as a legitimate aspiration to splendor but also as the obligation to continue its civilizing mission, it cannot and must not forget that, in possession of about two hundred million hectares of dispersed lands in the intertropical zone, it has the obligation of not delaying for its colonies what could contribute to its magnificence. Our doctors, as happens with those educated in British, German, Belgian, Dutch, French, and American schools, despite the superior learning of the professors, in finishing their courses ignore the living conditions in hot climates, with regard to individuals and their acclimatization, and to a general knowledge of the water, soil, diet, and of whatever is necessary for the productive functioning of medicine in the country, which is not what they were educated and prepared for. Ignorant of tropical hygiene and

[2] Bombarda, Miguel, "A creação d'uma escola de medicina colonial", Jornal da Sociedade de Ciências Médicas de Lisboa, 43, 1899, p. 350.

pathology, the sanitary maritime police, the hospital administration, (and) our doctors cannot practice their profession to the benefit of the inhabitants of the tropics.

The Lisbon School of Tropical Medicine (LSTM) was approved by the Minister of Maritime and Foreign Affairs, Antonio Teixeira de Sousa, and King Carlos I decreed the simultaneous creation of the School of Tropical Medicine and the Colonial Hospital, in accordance with the proposition of parliament, on the 24th of April 1902,[3]

The proposal to create these institutions wouldn't depend on public donation, as would happen in the United Kingdom and France, but would, in a modest way, prepare a building belonging to the Navy for the teaching of tropical medicine and for the establishment of a hospital for the workers and soldiers returning from overseas. Teaching would be split into three disciplines: pathology and general medicine, hygiene and climatology, and bacteriology and tropical parasitology, with the necessary laboratories. There would be a general course for the colonial doctors and lessons for the civilian workers, military personnel, missionaries, farmers, and businessmen. Instruction would be carried out by the overseas doctors and permanent Navy staff. The medical services of the hospital would be fulfilled by workers of the health bureau, without special remuneration. The auxiliary services would be undertaken by soldiers from the overseas health corps. Initial financing proposed an amount to sponsor the hospital that would be deducted in the form of taxes on the salaries of the personnel, from the registration fees and from the monies from the municipalities and municipal commissions of the colonies (1% of ordinary expenses). The cost of installation would be borne by the colonies.

This school hospital was, at first, provisionally located in an eighteenth-century building belonging to the Ministry of Maritime and Foreign

[3] Parliament Journal, 24.04.1902, 129–130; S.a, *Institute of Tropical Medicine—instructions for the 1951 academic year*, Publication n° 1, (Lisbon, 1951); João Fraga de Azevedo, *Fifty Years of Activity at the Institute of Tropical Medicine (1902–1952)*, (IHMT: Lisbon, 1952); Pedro Abranches, *The Institute of Hygiene and Tropical Medicine- a century of history 1902–2002*, (Celom: Lisbon, 2004).

Affairs,[4] the National Rope Factory (Cordoaria), where it remained until 1925.

The choice of this location, dictated by to financial and political constraints, is associated with the ideals of the empire in two ways: on the one hand, this space was used for the construction of canopies and ropes for sixteenth-century vessels during the Discoveries, and now it served to house this teaching hospital, with doctors and patients who traveled between the colonies and the empire's capital. It occupied a very peculiar geographical, physical, and medical space due to its refurbishment and was also an emblematic space, as the starting point for the conquest of the new world. It was symbolic for the country, for the desired clinical efficiency that was intended for a hospital specializing in tropical diseases, and for the cognitive space in which the new discipline of tropical medicine was being defined.

In 1902, the Portuguese medical community recognized the accomplishments of the 1901 mission, and the foundation of a specialized training center in tropical medicine in Lisbon by the Portuguese state was considered essential for the maintenance of the Empire. The school's objective was the "theoretical and practical teaching of tropical medicine" and the organization of "scientific missions to the overseas provinces and the foreign colonies". The hospital was intended for the "treatment of soldiers returning from overseas". The conditions were in place for recognizing that combating tropical diseases was the basis of colonization, and that scientific missions were the most effective tool for endemic mapping and for eradicating the "exotic pathologies" in the colonized territories, as affirmed by João Fraga de Azevedo, one of the directors of this institution, years later:[5]

> The participation of the School of Tropical Medicine in the colonization and civilization of overseas territory [...] was achieved through the

[4] For the foundation of the London School of Tropical Medicine, in 1899, Joseph Chamberlain, Minister of the British colonies, appealed for a donation of all those who could support colonial issues. The voluntary contribution produced 16,000 pounds. Similarly in Portugal, the Ministry of Colonies and Foreign Affairs donated 3500 pounds; the Ministry of India 1000 pounds, and the remainder of the donation being covered by individuals and financial and commercial companies, to support the creation of the School of Tropical Medicine and the Colonial Hospital, in Lisbon. Parliamentary Journal, 28.01.1902, p. 18.

[5] For a general understanding of the beginnings of tropical medicine in Portugal, see Fraga de Azevedo, op. cit. X, pp. 10–97.

intervention of the physicians who had specialized in teaching the profession at the School of Tropical Medicine, through study missions, and through dissemination and research work.

The Hospital and the School were institutions with a strong military influence. The management and clinical service of the hospital were the responsibility of the technical staff of the Health Department of the Directorate-General for Overseas and it was obliged to incorporate doctors from the Navy within its staff, with a minimum of two years of experience in the colonies. The first director of the school (as well as of the Colonial Hospital) was Ramada Curto, the longest-serving professor in the Navy, then head of the Health Services of the Ministry of Maritime and Foreign Affairs, and three other naval doctors. D. António and Ayres Kopke were entrusted with the organization of the school's first training course (Fig. 2.1).

Kopke, being a naval doctor and LSTM researcher, and a career doctor in Lisbon, soon stood out as a protagonist in the research on sleeping sickness. At the 15th International Congress of Medicine, held in Lisbon in 1906, he presented the first results of the research carried out on patients from the Colonial Hospital, using atoxyl in the treatment of the disease, concluding that it was effective in its first phase. These results were not only welcomed in the international scientific community by Robert Koch (1843–1910), Alphonse Laveran, Paul Ehrlich (1854–1915), and Patrick Manson but also catapulted him onto the international stage as a Portuguese representative and expert in tropical diseases. He took part in the first international conferences on sleeping sickness: the first three in London in 1907 and 1908, under the direction of Patrick Manson, and in 1925, on the invitation of the League of Nations; the fourth, in Paris in 1928, under the auspices of the League of Nations (Amaral, 2008).

Ramada Curto oversaw the theoretical and practical teaching of tropical medicine and equipped the laboratory at the Hospital de Luanda, stressing the importance of the study of tropical diseases in situ. D. António de Lancastre, the physician of the Portuguese Royal House, was the professor of exotic and clinical pathology, and in 1910, for ideological reasons, he was replaced by José de Magalhães on the advent of the Republic. Silva Teles, with his interests in colonial geography and anthropology, was an advocate for Portugal's geographical autonomy, and had experience in African colonies, particularly in Mozambique. He was responsible for teaching hygiene and climatology and assumed the role of school director,

Fig. 2.1 The parasitology class of Ayres Kopke at the School of Tropical Medicine. IHMT 0000350

replacing Ramada Curto in 1910. Ayres Kopke devoted himself to tropical bacteriology and parasitology. Due to the great number of students, the course program underwent its first reform in 1920 and was modified in 1932, thus contributing to the creation of the Institute of Tropical Medicine in 1935. In 33 years, the LSTM trained 708 physicians.

The Agenda of Scientific Missions in Portuguese Africa: Between Liberalism and Dictatorship

Considering the importance of sleeping sickness in the context of the colonization of Africa, and drawing upon Tilley's approach, two moments must be considered as the most decisive in the history of Portuguese tropical medicine: the eradication of sleeping sickness in Príncipe in 1914, during the First Republic (established in 1910); and the dispute over scientific authority between Kopke and Damas Mora, when the colonies' autonomy program by the institution of the High Commissioners' regime, followed by liberalism, gave rise to the dictatorial regime which began in 1926.

After completing the tropical medicine training course at the school, the qualified doctors applied to the Colonial Health Board, where they were hired as second-class doctors by either the national or foreign companies that operated in the colonies. While specialized training took place in Lisbon at a teaching and research institution created for this purpose, the cost of the clinical practice in Africa was at the expense of medical staff. These doctors 'in the bush' in Africa were almost entirely self-sufficient given their distance from their production center in the Portuguese capital. Caring for each patient became an adventure and a challenge for the doctors, and healthcare conditions were very precarious; the distances they were obliged to cover were enormous and dangerous, therefore, following up patients was a true odyssey (also given the resistance to European medicine and the doctors' constant displacement through the territory).[6]

Under the republican regime, the organization of healthcare services and medical assistance to Europeans and settlers in the colonies depended above all on how the administration was managed, which fluctuated between Lisbon's command and the political characteristics of each colonial High Commissioner. In this context, António Damas Mora and Bernardo Bruto da Costa deserve special mention, as both were responsible for improving healthcare conditions, particularly in Angola, São Tomé, and Príncipe, while in the service of the colonies' administration. Damas Mora was responsible for the introduction of a new approach in Angola in 1926, based on an integrated system of assistance to indigenous populations, as a way of making the colonization of Africa effective (Coghe, 2015, 2017; Amaral, 2018). Bruto da Costa was responsible for the implementation of major sanitation measures for the indigenous population in order to control the epidemiological mapping of the Portuguese colonies which began in Príncipe in 1911. These initiatives were essential for establishing Portuguese tropical medicine in the canons of African

[6] See, for example, the experiences reported by Calisto Martins Baptista, José de Brito Freire and Vasconcellos, Alberto de Sousa Maia Leitão and Annibal Correia Mendes (in Angola), António Damas Mora and Bruto da Costa (in Angola, Timor and Macau) or Carlos de Almeida (in the Portuguese Congo).

colonial health management, which were influenced and conditioned by the models and interests of European countries such as the United Kingdom, France, and Belgium.

The scientific missions organized by the LSTM targeted the most important overseas provinces most affected by sleeping sickness, which were Príncipe (1904, 1907, and 1911) and Mozambique (1910 and 1927).

In 1904, Ramada Curto nominated Ayres Kopke to study beriberi and sleeping sickness in Angola, and he continued his work in Lisbon with patients from the Colonial Hospital. In 1907, a new state mission was sent to Príncipe led by Correia Mendes and assisted by Arnaldo José Villela, Bernardo Francisco Bruto da Costa, and António Damas Mora, following news of a devastating epidemic in the territory. The objectives were more therapeutic, but the norms issued by the mission were not generally followed, so the mortality rate on the island of Príncipe continued to be alarming. In 1911, Miranda Guedes, Governor of the province, seconded Bruto da Costa to Príncipe on a mission directed by Correia Mendes and lasting until 1914. Organized under the patronage of the Geographical Society, the mission was sponsored by the Ministry of the Colonies and colonial companies interested in resolving this health problem. The action aimed to prevent the spread of this endemic disease on the island by eliminating the reservoirs inhabited by the parasite. It was not easy to convince the farmers to put the new measures into practice, despite the governor's cooperation, and even some of the local doctors did not believe in the effectiveness of the interventions recommended by Bruto da Costa.

The sanitary situation in Príncipe was intensified by the island's own economic development. With the increasing importation of slaves, the threat of the disease took on frightening dimensions to the point of endangering the habitability of the territory. This was because these individuals carried the disease and the tsetse flies also traveled on the same ships. These factors soon resulted in the 'pathologization' of the human environment. The situation called for the scientific occupation of the territory, which was also a consequence of the pressure exerted by the British, among them Clarence-Smith and William Cadbury, who denounced slavery in the territory and demanded measures for hired labor and the abolition of slavery (Amaral, 2018).

Bruto da Costa finished the LSTM course in 1905, sponsored by the state. According to the law, as the most outstanding student on the course, he was entitled to express a preference in his choice of destination in the overseas provinces. However, he was sent to São Tomé and Príncipe, the worst of all possible destinations, not only because of the climate but also because of the logistical conditions and the epidemics that raged there (beriberi, hemoglobinuric fever, sleeping sickness, etc.). He arrived there to be director of the bacteriological laboratory of São Tomé, where he started the first studies on the city's sanitation and became a pioneer in this regard by implementing a program for eradicating sleeping sickness on the island of Príncipe, which served as a model for other Portuguese colonies. This program, which began in 1911, not only achieved hygienic and sanitary goals but also political ones. As a result, it succeeded in "silencing" the interests of British cocoa traders. In five years, in collaboration with the Governor of the Province, the tsetse fly reservoirs were eliminated, and with them, the disease. The immigration of new workers to Príncipe was halted; the sick were quarantined, monitored, and treated with atoxyl; wild animals were slaughtered; marshy habitats were drained; the green security zones of the houses were extended; and the flies were captured by mechanical means with the collaboration of servants from the plantations. This original method of capturing tsetse flies was introduced in 1906 by Angelo Bulhões Maldonado, administrator of the Sundy Farm. Known as the "Maldonado method", it consisted of brigades of servants moving around the island, dressed in regulatory uniform (a white suit), which was covered with a fabric impregnated with a viscous substance to attract, catch, and kill the tsetse flies (Fig. 2.2). This method proved to be more effective than the British wooden traps and became one of the key elements in eliminating sleeping sickness on the island.

Príncipe was an ideal region for a scientific mission as far as the Portuguese were concerned. It had a limited territorial area and population, and most of the population was geographically stable, which allowed doctors to follow up patients for long periods. The sanitation measures proposed by Bruto da Costa led to the eradication of the disease. The report on the results of the disease eradication program was published in the *Archives of Hygiene and Exotic Pathology* in 1915 and was translated into English the following year (Wyllie, 1916). The claimed success of the program was publicized and applauded by the Portuguese authorities as they firmly believed that it placed the country, for the first time, at the forefront of European sanitary policies in Africa. Nevertheless, the British were unconvinced and traveled to Principe to confirm that the disease had been eradicated. In response to this, the British politician and banker William Ormsby-Gore also argued that "one thing is an island and another one is a continent" (Amaral, 2013).

Fig. 2.2 Brigade of servicemen with suits for capturing tsetse flies (of the *glossina* genus) on Príncipe Island. Relatório final da missão da doença do sono na ilha do Príncipe. Arquivos de Higiene e Patologia Exóticas, 1915: 111

The tension permeating British-Portuguese diplomacy since the 1890 Ultimatum resurfaced again. Together with a center versus periphery dynamic and the dispute over scientific authority, it affected political relations between the two countries.

In Mozambique, in 1909, sleeping sickness was spread by workers coming to Zambezia from British territories. In 1910, in the mission in Zambezia, Firmino Sant'Ana prepared a report for the Governor-General containing several proposals, including the prohibition, in Tete, of the employment of native Africans from British Central Africa and Northern Rhodesia, as well as the prohibition of the use of carriers from Northern Rhodesia and Nyasaland by Tete's commercial traders. On 13 February 1911, in a letter from the Directorate-General of Overseas Territories, a hypothesis was put forward that another fly from the *Glossina* family was spreading the disease in the Aruangua valley, a tributary of the Zambezi. The African World commission was appointed, and even though the Portuguese authorities were not consulted, they were advised that they would have to host the British mission, which was accompanied by Firmino Sant'Ana and José Pereira Lapa. Despite several complications in transportation and the hostile attitude of the British authorities in cooperating, a project to control sleeping sickness was created by the Governor of Mozambique and approved by the central government in 1914.

Three years later, Pereira Lapa (a student from the LSTM's 1906 cohort) drew up a report to continue improving the project and correct the most severe problems (for example, worker migration), which once again revealed the difficulty of imposing sanitary measures, even by the force of the law (Amaral, 2013).

The international conferences on sleeping sickness that were taking place in the meantime, and in which Ayres Kopke participated as a representative of the Portuguese school, led to a medical mission to Uganda (Entebbe) and Mozambique, led by Andrew Balfour. Belgium, the United Kingdom, and Portugal took part in this mission. Manuel Máximo Prates was the Portuguese doctor on this mission, which began in 1926, and was director of the bacteriological laboratory, which formed part of the Miguel Bombarda Hospital in Lourenço Marques. This collaboration resulted in the publication of a report presented to the League of Nations International Commission on Human Trypanosomiasis in response to the European program to control sleeping sickness that presupposed the blurring of geographical borders. The following year, the Portuguese mission led by Ayres Kopke was joined by two local doctors, Fontoura de Sequeira and Saraiva de Aguilar. They arrived in Mozambique with the aim of studying the geographical distribution of "African lethargy" in the equatorial and eastern regions of the continent, implementing the measures recommended by the League of Nations.

Although Angola was not one of the colonies most affected by the disease, it soon became engaged in mapping the distribution of the disease in the territory from 1901 onward. Between 1910 and 1915, João Baptista de Assunção Velho was responsible for sanitation in the Dondo region, making a medical census of the territory and looking for the existence of flies of the *Glossina* genus. In 1923, Damas Mora organized the first West African Tropical Medicine Congress in Luanda, over which he presided. This event became critical for the Angolan administration's agenda and had repercussions for other Portuguese colonies. These included a program for studying and controlling sleeping sickness, also reflecting the guidelines and interests expressed by international bodies, which promoted networked strategies and the bringing together of contributions from all countries in the territory. Carlos de Almeida was appointed to design a program to combat the disease in the Congo and in the Kwanza Region between 1923 and 1926.

António Damas Mora finished his medical training at the Lisbon Medical-Surgical School in 1901, was a disciple and personal friend of Ricardo Jorge, and enlisted in the Portuguese colonial staff by choice, given his interest in tropical medicine. He held various management positions in the healthcare systems of Angola, São Tomé and Príncipe, Macau and Timor. In all the positions he held, he stood out in the field of Hygiene and Public Health (or rather, Social Hygiene, as he preferred to say), dedicating much of his time to colonial administration and the organization of healthcare services in Portuguese overseas territories. He was also acting director of the Health Directorate of the Ministry of the Colonies, acting Governor-General of Angola, and director of the Institute of Hygiene and Tropical Medicine, from 1936 to 1939.

Given the diversity of the positions he held in such distant countries, aided by his time in the League of Nations at the invitation of Ludwig Rajchman, Damas Mora gained knowledge of the colonial health administrations of other countries such as France and Belgium, on which he built his own program of colonial administration and which he tried to implement in Angola. Two elements of his legacy deserve special attention: the first is the First Congress of Tropical Medicine of West Africa (Primeiro Congresso de Medicina Tropical da Africa Ocidental), organized in Luanda in 1923; the second is the Assistência Médica aos Indígenas project (AMI, Medical Assistance to Indigenous People), which he started in 1926.

The Luanda Congress was intended to bring together European countries with interests in Africa and thus establish a network of inter-imperial exchange in Central and West Africa, which was of great interest to Portugal. The proceedings of this congress were published in the *Revista Médica de Angola*, which was also the brainchild of Damas Mora and marked a significant boost to Portugal's status, as it had started in an overseas province.

The objective of the AMI was to respond to the demand for more humane treatment of Africans as per the rulings by international bodies. This model, similar to the French model implemented between 1890 and 1900, not only made it possible to map sleeping sickness in the province but also to develop community-based healthcare services, which only became a priority for the World Health Organization (WHO) after World War II.

The latent tension between the colonial administrations is not evident in the official publications since they were coordinated and directed by figures chosen by the Portuguese state, often in Lisbon. The epistolary correspondence, however, provides valuable insights into these aspects, notably between Damas Mora and Ricardo Jorge from 1926 to 1933, at a time when the former was Director of the Angolan Health Services and the latter, a member of the Hygiene Committee of the League of Nations.

The political and administrative tensions between Lisbon and the colonies were also rife in a dispute over scientific authority, given the disparity between the value of knowledge gained in the Lisbon laboratory and that of the School of Tropical Medicine in the African provinces that were envisaged as one vast laboratory (Amaral, 2018). In another letter from Damas Mora, this strained relationship became evident:[7]

> Contrary to the opinion of Mr. Kopke, who seems to have taken over the monopoly of the Portuguese fight against sleeping sickness, denying others the right to work in the same field, the Angolan health service has just received a resounding compliment from those in charge of the health services of the Belgian Congo, in the form of an opinion written in a record book of the sanitary region of North Kwanza (they considered Angola an example of perfection, in matters of organization).

Alongside the noteworthy achievements of Ayres Kopke, who pioneered many great scientific innovations and provided an indispensable link to the international network of experts in tropical medicine, those of Damas Mora are equally significant in the history of Portuguese tropical medicine. However, this dispute between Kopke and Mora, and the tensions between the republican project and the emergent dictatorship, is also illustrative of how sleeping sickness and the medical colonization of the African territory became more and more important in the Portuguese political scenario.

[7] Mora, António Damas. Letter to Ricardo Jorge in Lisbon, Cape Town, 8 December 1932 (personal archive; courtesy of Luiz Damas Mora). 1932.

Cambournac at WHO and the Route to Decolonization Under Salazar's Regime

The dictatorial regime led by Antonio Salazar after 1932 reinforced the nationalist and centralist agenda by conceiving of the colonial empire's program under the concept of "national unity". To Salazar's dictatorship, the Estado Novo (New State), the fight against this disease was not only a question of colonization but also an issue of international public health, as the international pressure that Portugal was subjected to practically forced the regime to redesign new sanitary policies as part of the colonial administrative apparatus.

Internationally, Portugal was at this moment trying to carve its own space within the WHO as a way of gaining credibility for its colonial project. Francisco Cambournac (1903–1994), a researcher and experienced malariologist at the Institute of Tropical Medicine, responsible for the malaria eradication program in Portugal with financial support from the Rockefeller Foundation, played a major role in this process. In 1946, he was invited to participate as a Portuguese delegate in the First International Health Conference in New York, and, therefore, to subscribe to the constitution of the WHO. The Portuguese presence in the WHO was necessary to ensure that the new international order involving the decolonization of overseas territories had Portugal as an ally. The participation of Portugal in the WHO inaugurated a new era in Portuguese tropical medicine: Cambournac became Regional Director for Africa (between 1954 and 1964) and placed several Portuguese researchers on the WHO list of experts in diseases such as malaria, trypanosomiasis, leprosy, viral diseases, and bilharziasis, and even in nutrition and health administration affairs. These doctors brought a new vision of social medicine to Portugal that would lead to the integration of the Institute of Tropical Medicine (founded in 1935) into the National School of Public Health in 1966.

The scientific missions carried out between 1942 and 1966 by the Institute of Tropical Medicine became increasingly relevant in the political discourse about the colonies in which European diseases, vectors, actors, institutions, and states crossed paths with the interests and strategies aimed at finding a global solution to pressing health issues and establishing a new order under the principle "health for all". In this context, the protection of the rights of all peoples became a golden rule, foreshadowing the end of European colonial rule in Africa. As an outcome of a reinterpretation of the Colonial Act of 1951, the missions organized by the Institute of

Tropical Medicine became permanent, with medical teams residing in the African colonies for several years, extending healthcare to the entire population. Some researchers from the Institute of Tropical Medicine tried to replicate the Lisbon Institute ofTropical Medicine in Angola, Mozambique, and Portuguese Guinea, the most important Portuguese African colonies, where there was a higher prevalence of sleeping sickness. This replication was a measure aimed at reviving the colonizing drive together with the agreement and involvement of local medical authorities.

The control of tropical diseases by healthcare systems established in Lisbon was becoming increasingly less relevant as the health services established in the African provinces became increasingly autonomous. However, whereas in Angola and Mozambique the campaigns against trypanosomiasis seemed to work effectively and with growing autonomy, in Guinea, the power of the Lisbon health authorities continued to be paramount (Havik, 2014). The trypanosomiasis mission in this colony was truly a State-Empire structure, and as such, tensions were high as its sanitary occupation was more effective. This was, therefore, used as a political instrument by the New State to keep the Portuguese African colonies under their control and influence as a matter of national unity (Fig. 2.3).

In his 10 years at the WHO, Cambournac boosted cooperation and increased the budget for WHO-AFRO. When the armed conflict in Portuguese Africa started in 1961 in Angola, criticisms against imperial policies increased. Colonial institutions were too intrusive to defend the interests of Africans, though at the same time they themselves were vocal in denouncing an anti-colonialist sentiment about forced labor practices, as did international observers on the ground.

The implementation of the Portuguese healthcare plan for its African colonies, which included not only curative medicine but also preventive medicine, in particular, was run from Geneva with the support of the WHO. It would have become much more attractive and decisive for the consolidation of Portuguese tropical medicine, both in mainland Portugal and overseas, if organizations such as The Commission for Technical Cooperation in Africa (CCTA), where Africans claimed positions in defense of their autonomy in public health policies, had not been so dysfunctional.

Fig. 2.3 Representation of the scientific mission of sleeping sickness in Guinea by a Guinean artist in 2014. (Courtesy of Luis Neves da Costa)

Some Final Concluding Remarks

Different ideologies and strategies shaped Portuguese colonial discourse and policy, as did the implementation of health programs (Strang, 1992), the common trait among them being the unity of the Portuguese Empire. However, these programs also had unforeseen outcomes, notably the fact that European public health intervention in Africa promoted decolonization and the desire for self-determination of these overseas colonies, with the Portuguese domains being no exception.

Tropical medicine was a key element in the dynamics of appropriation in the colonial territories and was a prominent issue in national and international disputes. This was evident in the controversy surrounding the discovery of sleeping sickness's etiological agent; the human trypanosomiasis eradication program in 1914 in Príncipe; and the program carried out by Damas Mora involving medical assistance to Africans in Angola in 1926. The presence of Cambournac as the Portuguese representative in the WHO culminated in the effective sanitary occupation of Guinea initiated in 1951. In Angola and Mozambique, a different route was followed as both colonies were able to create tropical medicine research institutes in 1955, modeled on the Lisbon-based Institute of Tropical Medicine.

The wounded pride resulting from the establishment of the "pink map" following the Berlin Conferences back in 1884–1885 had always overshadowed the Portuguese medical occupation of its African colonies, causing different reactions at various times. However, this circumstance also gave tropical medicine a prominent role in sustaining the Portuguese empire.

In addition, ascribing the label of experts in colonial medicine to the doctors serving in the Portuguese African territories resulted in better working conditions for them, which meant they could respond more effectively to local medical problems than the doctors working in mainland Portugal. Since the early part of the twentieth century, the etiological agent for most of the "tropical diseases" and disease transmission routes were already known. There was, however, an impending need to build up the empire's infrastructures without undermining the fight against the propagation of some vector-borne diseases, which could, in turn, jeopardize the success of the imperial project. At the same time, it was necessary to comply with international guidelines emanating from organizations such as the League of Nations or the WHO, which advocated for the defense of the rights of African people.

The cult of the microbe spread to the parasite, and the insect pervaded the political sphere. In the networks of biopower, it is possible to reinterpret the "consolidation" of the Portuguese empire during the twentieth century by relating it to the importance of Portuguese tropical medicine. This reveals the continuities, tensions, and discontinuities, both on mainland Portugal and overseas, as well as the disputes over authority in the national and international scene, around "true" tropical diseases, of which sleeping sickness is an example.

References

Abranches, P. (2004). *The Institute of Hygiene and Tropical Medicine- A Century of History 1902–2002*. Celom.

Alexandre, V. (2000). *Vellho Brasil, Novas Áfricas – Portugal e o Império (1808–1975)*. Edições Afrontamento.

Alexandre, V., & Dias, J. (1998). *Nova História da Expansão Portuguesa. O Império africano 1825–1890* (Vol. X). Estampa.

Amaral, I. (2006). The Emergence of New Scientific Disciplines in Portuguese Medicine: Marck Athias's Histophysiology Research School, Lisbon (1897–1946). *Annals of Science, 63*(1), 85–110.

Amaral, I. (2008). Building Tropical Medicine in Portugal – The Lisbon School of Tropical Medicine and the Colonial Hospital (1902–1935). *Dynamis, 28*, 299–336. https://doi.org/10.4321/s0211-95362008000100013

Amaral, I. (2012). Bacteria or Parasite? The Controversy over the Etiology of Sleeping Sickness and the Portuguese Participation, 1898–1904. *Historia, Ciencias, Saude – Manguinhos, 19*(4). https://doi.org/10.1590/S0104-59702012005000004

Amaral, I. (2013). As Missões Científicas Nas Colónias e a Escola de Medicina Tropical (1902–1935). *Boletim da Sociedade de Geografia de Lisboa, 636*, 53–60.

Amaral, I. (2018). Medicina Tropical e Ambiente em Perspectiva: Reconstituindo o Puzzle da Erradicação da Doença do Sono na Ilha do Príncipe, Em 1914. *Fronteiras- Journal of Social, Technological and Environmental Science, 7*(2), 64–82. https://doi.org/10.21664/2238-8869.2018v7i2.p64-82

Amaral, I. (2020). Portuguese Tropical Medicine and International Networks: A Global Health Governance Project in the 20th Century. *Vesalius, Journal of the International Society for the History of Medicine, XXVI*(e-supplement), 247–262.

Bombarda, M. (1899). A creação d'uma escola de medicina colonial. *Jornal da Sociedade de Ciências Médicas de Lisboa, 43*, 350.

Coghe, S. (2015). Inter-Imperial Learning and African Health Care in Portuguese Angola in the Interwar Period. *Social History of Medicine, 28*(1), 134–154.

Coghe, S. (2017). Sleeping Sickness Control and the Transnational Politics of Mass Chemoprophylaxis in Portuguese Colonial Africa. *Portuguese Studies Review, 25*(1), 57–89.

Curtin, P. (1998). *Disease and Empire: The Health of European Troops in the Conquest of Africa*. Cambridge University Press.

Doença do Somno. (1900). Representação dirigida ao governo pela Sociedade das Sciencias Medicas. *Jornal da Sociedade das Sciências Médicas de Lisboa, LXIV, 7 a*(12), 268–269.

Fanon, F. (2020). *Medicina e colonialismo*. Editora Terra sem Amos.

Fraga de Azevedo, J. (1952). *Fifty Years of Activity at the Institute of Tropical Medicine (1902–1952)*. IHMT.

Havik, P. (2014). Public Health and Tropical Modernity: The Combat Against Sleeping Sickness in Portuguese Guinea, 1945–1974. *Historia, Ciencias, Saude – Manguinhos, 20*(2), 653–673.

Havik, P., Jeronimo, M., & Castelo, C. (2017). Historical Trajectories of the Third Portuguese Empire: Reexamining the Dynamics of Imperial Rule and Colonial Societies (1900–1975). *Portuguese Studies Review, 25*(2), 3–13.

Headrick, D. (1981). *The Tools of Empire: Technology and European Imperialism in the Nineteenth Century.* Oxford University Press.

Lemaine, G., Macleod, R., Mulkay, M., & Weingart, P. (Eds.). (1976). *Perpectives on the Emergence of Scientific Disciplines.* The Hague and Maison des Sciences de l'Homme.

Lyons, M. (1992). *The Colonial Disease: A Social History of Sleeping Sickness in Northern Zaire, 1900–1940.* Cambridge University Press.

Mora, A. D. (1932). Letter to Ricardo Jorge in Lisbon, Cape Town, 8 Dec. 1932 (Personal Archive; Courtesy of Luiz Damas Mora).

Neill, D. J. (2012). *Networks in Tropical Medicine: Internationalism, Colonialism and the Rise of a Medical Speciality.* Stanford University Press.

Oliveira e Costa, J. (coord.). (2014). *História da Expansão e do Império Português.* A Esfera dos Livros.

Parliament Journal, 24.04.1902.

Parliamentary Journal, 28.01.1902.

Relatórios enviados ao Ministério da Marinha pela missão scientifica nomeada por portaria de 21 de fevereiro de 1901. Imprensa de Líbano da Silva. 1901.

S.a, Institute of Tropical Medicine – Instructions for 1951 Academic Year.

Sabbe-Clarke, E. E., Bradley, D. J., & Kirkwood, K. (Eds.). (1980). *Health in Tropical Africa During the Colonial Period.* Clarendon Press. Said, E. 1993.

Shapiro, F. (1983). *Medicine in the Service of Colonialism: Medical Care in Portuguese Africa (1885–1974).* PhD Dissertation, University of California.

Steverding, D. (2008). The History of African Trypanosomiasis. *Parasites Vectors, 1*, 3. https://doi.org/10.1186/1756-3305-1-3

Strang, D. (1992). The Inner Incompatibility of Empire and Nation: Popular Sovereignty and Decolonization. *Sociological Perspectives, 35*(2), 367–384.

Tilley, H. (Ed.). (2011). *Africa as Living Laboratory: Empire, Development, and the Problem of Scientific Knowledge, 1870–1950.* University of Chicago Press.

Worboys, M. (1976). The Emergence of Tropical Medicine: A Study in the Establishment of a Scientific Specialty. In G. Lemaine, R. MacLeod, & M. Mulkay (Eds.), *Perspectives on the Emergence of Scientific Disciplines* (pp. 75–98). The Hague.

Wyllie, J. (1916). *Sleeping Sickness: A Record of Four Years' War Against It in Principe, Portuguese West Africa.* Baillière, Tindall and Cox.

CHAPTER 3

Dutch Colonial Medicine and Empire Building in the Tropics: The Cases of Leprosy and Drug Use in the Dutch East and West Indies Compared

Stephen Snelders

Introduction

In this chapter, I will focus on colonial policies around leprosy and drug use/abuse/addiction in the Dutch East and West Indies (in particular, in what are now the nations of Indonesia and Suriname) as constituent parts of Dutch colonial and imperial policies. I recognize, of course, that leprosy and drug use were very different problems both from the perspective of public health and of public order. I find it the more striking then that the framing and management of both these problems, as different as they are in nature and character, show important similarities in colonial societies.

S. Snelders (✉)
Freudenthal Institute/History and Philosophy of Science, Utrecht University, Utrecht, The Netherlands
e-mail: s.a.m.snelders@uu.nl

M. Capocci, D. Cozzoli (eds.), *Empire, Nation-building, and the Age of Tropical Medicine, 1885–1960*, Medicine and Biomedical Sciences in Modern History,
https://doi.org/10.1007/978-3-031-38805-7_3

In the Dutch colonial empire, leprosy was framed since around 1750 as a disease connected to notions of racial inferiority of the majority of its sufferers, and as a threat to the functioning of the empire. This translated into far-reaching policies of monitoring of the population, and isolation and segregation of sufferers, that peaked between 1830 and the Second World War. Drug use became (internationally) framed as a problem of 'racially inferior' population groups in the colonies in the course of the nineteenth century, leading to the first anti-drug policies on Dutch territories by the 1890s, and from then onwards to Dutch participation in the international regulatory drug regime headed by the League of Nations and the United States.

The term 'framing' is used here in the sense as developed by medical historian Charles Rosenberg: the fashioning of explanatory schemes for particular diseases and disorders. In Rosenberg's influential perspective, these schemes, and the very process of defining diseases, whatever their somatic basis, are affected by values, attitudes, and status relationships in the larger culture (Rosenberg, 1989). In this chapter, we will look at processes of 'colonial' framing, or the framing of diseases in (Dutch) colonial society. Both leprosy and drug abuse/addiction were framed as socially threatening diseases in processes that were driven by local developments in these colonial societies, but also stood within a wider colonial and imperial context. Colonialism, as it developed in the Dutch empire, increasingly meant the need for more control of geographical territories, as well as of global migration movements. Policies evolving in colonial territories aimed at controlling colonial non-white populations. Colonial framing, connecting diseases with 'inferior' (and subjected) 'races', was a central part of these policies, as well as of the process of understanding and managing the diseases, resulting in stereotyping and stigmatizing leprosy sufferers and drug users. 'Othering' these groups contributed in essential ways to the formation and development of the colonial state.

The comparison of the framing of drug users to another process of framing of a disease that took place at the onset of imperialism, that of the leprosy sufferer, might at first seem surprising, but is not a new one. In the 1960s, when a large drug-using culture expanded and prohibition became a contested issue, vocal proponents of this drug culture formulated the notion of drug users as victims of society, criminalized and stigmatized. Drug users were, according to the influential sociological theory of Howard Becker, made into *outsiders* (Becker, 1963). Users compared themselves to persecuted medieval witches and lepers. Michel Foucault traced in his *Histoire de la folie* (*Madness and Civilization*, 1961) the genealogy of exclusion to the framing of leprosy in medieval Europe. He suggested that the stigma of lepers survived long after leprosy itself had become extinct in Europe and the leprosy asylums had emptied. However, 'the values and images connected to the leprosy sufferer continued to exist' (Foucault, 1995, pp. 17–18). The games of exclusion were played again, but now with the insane. Thomas Szasz (1985) in his turn stated that this stigma had been transferred to drug users as well.

Whatever the pros and cons for the theses of Foucault and Szasz may be will not be the subject of this chapter. My investigation will be more limited and addresses the question what a comparison of Dutch colonial polices around leprosy and drug use will tell us about the construction of the Dutch colonial empire. It will also address the question between the relationship of framing and policies.

Suriname

The two colonial territories I will focus on (Suriname or Dutch Guiana on the mainland of South America, and the Dutch East Indies or present-day Indonesia) were vastly different in size, population numbers, system of political control, and extent of contacts between colonizers and colonized. Suriname was a Dutch construct. After the Dutch took control in 1667, they created a plantation society in which the vast majority of the population were 'imported' enslaved Africans who had to be controlled. By the 1780s, there were 1500 Europeans, 33,000 enslaved Africans, and 500 multiracial people in the colony (Snelders, 2017, p. 23). With the exception of the Maroons and their descendants in the interior, who after prolonged warfare had gained autonomy, people in Suriname lived under direct Dutch control. Until 1795, Suriname was ruled by the Society of Suriname (a private company with the city of Amsterdam and the Dutch West India Company among its shareholders), while after the revolutionary and Napoleonic period, it was under control of the Kingdom of the Netherlands. Slavery was only abolished as late as 1863. Since then, and until after the Second World War, Suriname was considered part of the Netherlands, where people needed to learn to be Dutch citizens. In the view of the Dutch government, Suriname constituted 'an overseas Dutch territory the population of which must be granted all the privileges and liberties which have been through the ages the heritage of every Netherlands citizen' (van Lier, 1971, p. 190). Population increased from almost 53,000 to 69,000 in 1898 (of which only 633 were Europeans). Production of Suriname's chief exports, sugar and its by-products such as molasses and rum, shifted to large-scale plantations, and the increase in the population was due to the recruitment and immigration of indentured labourers from China at first, and later from British India and Java in the Dutch East Indies. This population had, however, to be taught to be Dutch citizens, to accommodate into the dominant culture. African religious practices were constituted as criminal offences in 1874, and compulsory education in the Dutch language was introduced in 1876 (van Lier, 1971, pp. 191–193). We will see how the reconfiguration of leprosy politics and the introduction of innovative drug regulations, both starting in the 1890s, were closely connected to these developments.

Dutch East Indies

The Dutch East Indies were much larger but also much more heterogeneous than Suriname. The archipelago consisted of a great variety of ethnicities, cultures, languages, and societies that to a large extent remained autonomous. By 1890, the population was estimated at about 43 million. Given the complex power structures and the sheer size of the Dutch East Indies, it was difficult for the Dutch to gain control. As a consequence, Dutch territorial rule was much more limited than in Suriname. Outside of Java (the most important island) and the Moluccas, the Dutch had very little power in the archipelago until the nineteenth century. But even on Java, the Dutch behaved as a 'reluctant imperialist', mainly concerned with maintaining their influence on trade and keeping the balance of power between the different Javanese principalities (Gaastra, 2002, p. 64). In the nineteenth century, a system of indirect rule was introduced in the East Indies, in which Javanese communities produced coffee, tea, and sugar for the Dutch. The resulting political structure can be characterized as an association of Dutch and Javanese elites. Elsewhere in the archipelago, Dutch political power remained mostly nominal and in fact non-existent. Only by the end of the nineteenth century did the modern colonial state start to extend its sovereignty to the other islands by military and other means (van Doorn, 1994; van den Doel, 1996; de Jong, 1998). Policy interventions regarding leprosy and drug use were, therefore, never as encompassing and invasive as in Suriname, and they took different forms.

The 'Imperial Danger': Leprosy in the Global Context

These policies have to be placed in a more international comparative context. Leprosy as a perceived 'danger to the empire' is by historians mostly associated with the end of the nineteenth century, the Age of Imperialism, and the global onset of scientific racism and Social Darwinism (cf. Bashford, 2004). *Leprosy, an Imperial Danger* is in fact the title of a tract published in Britain in 1889 (Wright, 1889). At that time, leprosy came to be regarded as an 'imperial disease', moving throughout the European colonial empires through migration of people and circulation of goods, and affecting white people in the process. In 1989, Zachary Gussow concluded in his study of American leprosy politics in an international context:

> [...] by the nineteenth century [leprosy] had reappeared and by the end of the century had caused Western nations to panic. During the period of nineteenth-century imperialism, the disease was discovered to be hyperendemic in those parts of the world that western nations were annexing and colonizing. The discovery of leprosy in the colonial world, and the excitement in the 1860s generated by the announcement of an epidemic in Hawaii, revived Western concerns about a disease that otherwise remained but a memory. (Gussow, 1989, p. 19)

A former medical officer of the southern American Confederacy wrote in 1887: 'At the present day Louisiana is threatened with an influx of Chinese and Malays, with filth, rice and leprous diseases [...] patriots should contemplate with dread the overflow of their country by the unprincipled, vicious and leprous hordes of Africa' (Gussow, 1989, p. 57). Leprosy seemed to threaten the United States, and in the 1880s measures were taken to prevent Chinese leprosy sufferers to come into the country. Attention was directed to white leprosy sufferers as well. Before 1880, they seemed to have been more or less left in peace in the United States, but now doctors actively went searching for carriers of the disease. In 1917, Congress decided on the creation of a national leprosy asylum to isolate the sufferers (three years, one notices, after the passing of the Harrison Narcotics Act that regulated the opiates and cocaine market). Other Anglophone countries such as Australia, New Zealand, and Hawaii took their own measures against the Chinese immigrants from the 1880s onward. A Scottish physician working in Hong Kong, James Cantlie, suggested in an 1897 report that all migrants should be examined, and sufferers expelled (Edmond, 2006, pp. 125–170).

Gussow related the 'rediscovery' and renewed fears of leprosy to anxieties about Chinese immigration and the endangerment of 'American-ness'. But interestingly, similar processes were at work in China itself and connected to concepts and desires of nation-building. To the Chinese nationalists behind the anti-Manchu revolution of 1912, leprosy became a sign of the undisciplined body, a matter of shame for China, necessitating isolation and public health policies (Leung, 2008).

Leprosy in Suriname

However, similar framings of leprosy sufferers had taken place in the Caribbean earlier, in the eighteenth century. The changing disease environment and the close proximity to slaves of African descent prompted inquiries in health and disease of the non-white population in the Caribbean much earlier than in Asia. By the later eighteenth century, the topic of what Londa Schiebinger has called the 'anatomy of difference' between races was widely debated among the scientists and savants of Europe. Explanations for these 'differences' ranged between environmentalism and hereditarianism, including combinations of both (Schiebinger, 1989/1990). While in Europe this was of more theoretical concern, in the colonies, the question of why and to what extent different races were prey to different diseases was of eminently practical concern and became closely connected with issues of constructing colonial societies and ordering the lives of the vast majority of the population of these societies: the black enslaved. As Sean Quinian wrote in his study of the French colonies, the ultimate distinction between the races was located in the amount of self-control a male European could exert in order to regulate his functioning in accordance with the environment. 'In a sense, the diseased body became the ultimate signifier of not just the pathological milieu but the total lack of physical self-control exercised by the European individual', wrote Quinian (1996, p. 120).

Leprosy became in Suriname a key disease in which racial stereotyping and 'Othering' led to far-reaching policies of monitoring and compulsory segregation. This process started in the 1750s, when an unknown disease among slaves was observed, the signs of which reminded doctors of descriptions of leprosy in medical textbooks. It was called 'boasie', after the name of the place in Africa where the disease was believed to originate. In the eighteenth century, many European doctors and laymen regarded the disease as highly contagious. In Suriname, healthy persons were advised to stay away from sufferers, not to enter their dwellings, not to touch them, nor even breathe the same air. To many slaveholders, leprosy seemed to endanger the health of their slave labour force and with it the proper functioning of the Surinamese economy. Profits would be seriously at risk if changes in the disease environment were not met by medical intervention. The colonial framing of leprosy that followed decisively influenced dealings with leprosy in Suriname. In 1790, the colonial government issued even stricter measures, introducing compulsory segregation that forced slave owners to report possible leprosy cases among their slaves to the *Collegium Medicum*. When this board of medical supervisors diagnosed a slave as being affected by leprosy, the affected slave would have to be moved to a special leprosy colony called 'Prevention' (*Voorzorg*), in 1824 replaced by the Batavia asylum. (for a fuller analysis, see Snelders, 2017, pp. 21–30).

The edict of 1790 can be considered as a codification of the colonial framing of leprosy that was formed during the preceding decades. The German slave doctor Godfried Wilhelm Schilling had played an all-important role in this process (Snelders, 2013). As one of the few physicians working in the colony of Suriname, and one who conducted medical examinations of newly arrived slaves, Schilling supplied the scientific 'evidence' and underpinnings of the repressive policy of isolation and segregation based on a humoral understanding of the disease. He claimed that leprosy was caused by a 'special substance', 'a certain poison' that was brought by slaves from Africa to the Americas and that could become virulent when climate or inadequate diet had weakened a person's constitution. Race was an essential category to Schilling in understanding the aetiology and epidemiology of leprosy, and important in determining the measures that needed to be taken. He emphatically pointed out that Africans were the carriers of the disease. In doing so, he justified the edicts aiming to isolate leprosy sufferers. Schilling felt that whenever Europeans had physical contact with Africans, they were in danger of being contaminated. There were distinct moral connotations to this view: Africans were considered prone to contract the disease because they lacked the self-control and level of civilization needed to withstand the leprosy poison. At first glance, it seems that Schilling put the blame entirely on the Africans. However, in his view European males were in danger of contracting leprosy as well, in particular when they had sexual intercourse with the enslaved. The moment they lost their self-control, stooping to the level of a race that was held to be inferior, they were prone to contract leprosy as well. The more one was ruled by lust, the bigger the risk of leprosy spreading from the enslaved to master. Thus, moral, social, and medical degeneration went hand in hand, endangering European dominance in a society largely consisting of Africans. Schilling framed leprosy as a disease of African origin, tainted with negative racial and sexual connotations, caused by unhealthy living conditions, and related to low sexual standards. In his colonial gaze, the condition of leprosy was intimately connected to the identity of the African sufferer, both bodily and psychologically. His framing of leprosy was fully in line with the general views of the Dutch in Suriname. They had come to see leprosy as a problem of geography and race, as a threat to the functioning of the slave economy, and ultimately as a danger to European dominance (Schilling, 1771, pp. 4–40; see also Snelders, 2013).

Doctors in early nineteenth-century Suriname routinely embraced Schilling's sexualized and racialized discourse on leprosy. They complained about the supposed fatalism of Africans and leprosy sufferers, and expressed fear for a 'return' of the disease to Europe (e.g., van Hasselaar, 1835). In 1828, a major change in Dutch slave legislation necessitated new leprosy policies. Before that year, the enslaved had had the legal status of mere objects, owned by their masters, who could do with them whatever they liked (up to a point). In 1828, their legal status was changed from object to person, making new legislation on leprosy necessary. For the next 100 years, the leprosy edict of 1830 was to serve as the legal framework for leprosy policies. It was built on the foundations of the edict of 1790 that had first formulated the principle of compulsory segregation. Extensive powers of criminal investigation to track down suspected sufferers were given to the police of Paramaribo, the capital of the colony and the only settlement with a significant population outside of the plantations. The police were supposed to make a general visitation of all dwellings in Paramaribo once every three years, starting in 1831. Before they were put up for auction, the enslaved were to be examined by the city physician. Rewards were promised for every enslaved leprosy sufferer that was brought before a medical Committee of Investigation (*Commissie van Onderzoek*). The Committee directed its 'colonial gaze' especially (although not exclusively) at the enslaved brought before them by their owners (Snelders, 2017, pp. 43–77). Since almost all people who were examined or diagnosed as infected were enslaved, it can be calculated that almost 2% of the total slave population in 1837 had been confronted with the 'colonial gaze' on leprosy, something that did not occur in the case of any other disease.

Since the mid-1850s, there was a decline in the intensive monitoring and segregation of the enslaved because of a growing awareness of the impending abolition of slavery. Slave owners were unwilling to report leprosy among their enslaved workforce because this implied losing financial compensation when Emancipation would come into effect. The end of slavery did not, however, mean the end of compulsory segregation in Suriname. The social and cultural heritage of slavery continued to have its impact on dealings with leprosy. Post-Emancipation Suriname remained permeated by what has been called a *culture of domination*. The colonial state, social and religious organizations, and employers attempted to exercise authoritarian control (Hoefte, 2014). At the same time, fear of leprosy sufferers remained. By 1930, 80% of those diagnosed with leprosy were still Afro-Surinamese (Lampe & Simons, 1929). Numbers of segregated sufferers rose again after three new modernized asylums had been opened in the 1890s to replace Batavia (Snelders, 2017, pp. 161–198). In the context of international fears about leprosy as an imperial danger, the Dutch revived and revivified a colonial legacy from the period of slavery.

Leprosy in the Dutch East Indies

Colonial leprosy policies in Suriname were intrinsically linked with perceived notions of racial inferiority and needs to establish distance between colonizers and colonized. This view is corroborated with a comparison with leprosy policies in the Dutch East Indies. Distance between the Dutch and their subjects was in general larger in the East Indies than in Suriname. As a consequence, in the Dutch East Indies leprosy was never considered the danger it was perceived to be in Suriname. Medical services, even in a relatively 'European' town like the capital Batavia (now Jakarta), were focused on the military, leaving little personnel for civil health services. In Batavia, public health was not considered a major concern of government and attempts to improve living conditions for non-Europeans were virtually non-existent. Only a minority of the Dutch colonial administrators in Java advocated complete segregation of leprosy sufferers. The others either chose to leave policy to the discretion of the local Javanese authorities—which was in line with the principle of indirect rule—or rejected the idea of segregation altogether. It is significant that while in Suriname the number of sufferers was meticulously counted, similar statistics were lacking for the Dutch East Indies. Estimations made after 1900—running into tens of thousands, even 150,000 sufferers—had emotional rather than real value (van Driel, 1922). They were an indication of a rather abstract popular fear in the absence of concrete leprosy policies. What did cause anxiety was the arrival of Chinese labourers in the Dutch East Indies in the 1890s. In the public media the Chinese—who had traditionally been viewed with suspicion by the Dutch—were held responsible for an increase in the transmission of leprosy. The arrival of Chinese labour migrants stimulated constructions of their 'Otherness', and leprosy became associated with the Chinese and their supposed racial inferiority, as happened in other European colonies and in the United States (see above). Still, even the influx of Chinese did not lead to a change in general policies, let alone to compulsory segregation. Unlike the situation in Suriname, the physical distance between colonial rulers and ruled was too large to invest in large-scale measures. There was, however, one notable exception to this rule: the Deli tobacco plantations on the island of Sumatra, a place where labour conditions were very similar to those on the slave plantations in Suriname. This, in turn, led to a very similar response to leprosy and the forced segregation of (Chinese) sufferers in the context of labour policies (van Bergen, 2018; Snelders et al, 2021).

Orientalization of Drug Use

By that time another imperial danger was perceived by the colonial rulers: drug use among coloured labourers, endangering work efficiency. These perceptions were built upon an international 'orientalization' of drug use that increasingly dominated discourses in the course of the nineteenth century, leading to a new concept and object of intervention, the illicit drug. British writer Thomas de Quincey and his *Confessions of an English Opium Eater* (1821), maybe the earliest and certainly the most well-known example of the oriental framing of opium use, is exemplary. As Alethea Hayer wrote, De Quincey as a European opium user who had free access to the drug 'did not break any law; public opinion was not against him or even focused on him' and supplies of his drug were cheap and easily available (Hayther, in de Quincey, 1986, p. 7). However, De Quincey did produce an exemplary cultural image of the use of his favourite drug (laudanum, an opium tincture) as a threat from oriental regions. In his 1991 study of De Quincey, John Barrell claims that De Quincey was terrorized by the fear of an unending and interlinked chain of infections from the East, which threatened to enter his system and to overthrow it, leaving him visibly and permanently 'compromised' and orientalized. The Orient here was actually an undifferentiated mixture of Chinese, Indian, Egyptian, and other images forming—in the words of Barrell—one 'pathological identity'. 'The Orient [for De Quincey] is the place of a malign, a luxuriant or virulent productivity, a breeding ground of images of the non-human, or of the no less terrifyingly half-human, which cannot be exterminated, except at the cost of exterminating one's self, and which cannot be kept back beyond the various Maginot lines [...] that De Quincey attempts to defend against the 'horrid enemy' from Asia' (Barrell, 1991, p. 19). To quote De Quincey (1821) himself: 'I was kissed, with cancerous kisses, by crocodiles; and laid, confounded with all unutterable slimy things, amongst reeds and Nilotic mud'.

As Alan Bewell points out in his *Romanticism and Colonial Disease* (Bewell, 1999), this threat from the Orient received a powerful expression in De Quincey's account of the Malay that one day knocked at his door. The Malay was a fellow opium-eater, and despite his cultural otherness, he was akin to De Quincey himself. However, he was also a sexual threat to the white race, as typified in his encounter with a beautiful young English girl who had never before seen anything Asian.

In the orientalization of opium use, the user loses the essential qualities considered necessary for white men to deal with oriental threats (and the subjected people in the colonies): self-control and work ethic. De Quincey later in his life appears to succumb to the inactivity and torpor of the opium-eater. He himself was never an outcast of society. But his oriental framing of his drug use, with the associations of loss of self-control and of identity, confounded in a slimy pit, points to the future. It raises the kind of social fears that turned up in the European colonial empires: of the white man going native, losing the very core essence of his racial identity and becoming a non-European himself. This is a not only a danger for individuals but by extension for European rule over other races itself. As Bewell has shown, the drug problem of De Quincey is not a national English, but a colonial problem. In the colonies the danger of the Chinese, by then reconfigured as the 'yellow peril' to the West, appears again. By the end of the nineteenth century, as Howard Padwa in his study *Social Poison* (Padwa, 2012) discussed, the association of opium use and Chinese was especially established as a threat to the empire.

Similar to the case with leprosy, the Chinese, however, had their own problems with drug use. Activists struggling for a more powerful China and the end of the 'feudal' Manchu dynasty had casted opium use as a vicious habit brought in by the Western barbarians to undermine the celestial empire (a notion that is rejected by Dikötter et al., 2016 in their history of Chinese drug culture). In China, the anti-opium offensive connected to the Opium Wars, and later became related to the rise of a Chinese bourgeoisie and Chinese nationalism (Madancy, 2004). To the nationalists, opium poisoned the nation, especially after the lost war against Japan of 1895 (Dikötter et al., 2016, p. 108).

In both East and West in the 1890s, drug use, as leprosy, has become more than a disease or disorder. Gussow made the important observation that a disease (or a disorder) in itself does not have to be perceived as socially threatening. 'The condition must threaten a class of citizens, a way of life; it must arouse public concern; above all, it must give rise to action. It must become a social problem' (Gussow, 1989, p. 4). This had become the case with leprosy and drug use. And just as with the former, the Dutch colonial rulers translated their views on the social problem of drug use in policies.

Drug Regulations in the Dutch East Indies

By 1900, drug regulations were in place in Dutch colonies—actually, even before these kinds of regulations were promulgated in the Netherlands itself and affected the views of the metropolitan Dutch towards opium and other substances. Opium, after all, had always been an indispensable ingredient of European medicine since at least Antiquity (Booth, 1996). In the seventeenth century, at the beginning of the Dutch colonial period, the medical use of opium was stimulated by the influence of Paracelsianism, so that the drug was common in Dutch pharmacies (Douwes, 2020). In the eighteenth century, Dutch newspapers occasionally referenced to opium abuse (for instance, the administration of opium as a kind of date rape drug, or overdoses and suicides) but the dominant representation was of opium as a medicine, even when it was brought in from exotic places in Asia, more particularly, Turkey (Snelders, in preparation). In the Dutch colonies, however, a completely different narrative started to develop, and in the twentieth century, it would become the dominant narrative on opium and its users. Addiction to opium became a 'colonial disease': a disease connected to global and colonial expansion and exchanges, one that does not stay in the colonies but threatens the homelands of the colonial rulers (Bewell, 1999, p. XV). The paradox here was that the Dutch themselves traded in opium. Colonial drug regulation after all developed out of the economic motives of the opium trade: the defence of the opium monopoly of the Dutch East India Company (VOC), and since the 1800s of the colonial state in the Dutch East Indies against interlopers and smugglers. The dialectics of a regulatory regime were at place here, confronting Dutch colonial strategies with smuggler's tactics (Snelders, 2021, pp. 25–26, 44–46).

Around 1800 the prevalence of smuggling became paradoxically an argument *against* a total ban of opium use in the Indonesian archipelago, since such a ban was thought to lead to an increase of the illegal economy. But by that time Dutch politicians had reconceptualized opium use. Originally primarily regarded as a medical practice among the people of the archipelago (as it was in the Dutch Republic), it was by then imagined as a 'slow-working poison' endangering millions of people. Opium had once been associated with Sumatran warriors going on a rampage but became more and more related to orientalized images of passivity as discussed above, its use leading to dependence, diminishing sexual appetites, and reduced work performances by agricultural labourers. In this perspective, on the one hand opium use was a threat to the colonial economy, but even when demonized it was on the other hand an important source of colonial profits. The answer to this ambiguity was keeping the state's trade monopoly intact and even stimulating the trade's growth, while at the same time taking public health measures against abuse. This view continued to lie at the basis of Dutch colonial opium policies until the Japanese occupation of Indonesia in 1942. It led to the introduction of the so-called *Opiumregie* (Opium regime) that by 1914 was in force throughout the East Indies. Besides entailing a state monopoly on the import of crude opium, it also made the production of *rookopium* (smokable opium) a state monopoly. Crude opium, derived from the drained and dried juice of poppy heads and the base for the preparation of smokable opium, was imported from British India. The distribution of the prepared product was arranged through a system of concessions, with differentiations in the availability of the drug in particular regions and for specific population groups. In this way, the Dutch government hoped to increase profits and at the same time exclude the Chinese intermediate traders, who until then had bought the rights for the distribution of the drug from the state. The new system also facilitated the fight against opium smugglers, since the new standard product of the colonial state factory was easily discernible from illegal products. Not all users were, however, pleased with state opium and many preferred opium of other taste and quality. A police task force was set up to fight the illegal opium trade but had little success. Illegal opium and later opium-based drugs such as heroin and morphine were smuggled in from Malaysia and China. Arguments for the Opiumregie were, however, not limited to economic or public security issues. The Dutch government claimed that increased control of the whole chain of opium import, production, and distribution would lead to a reduction of opium abuse and dependence among the subject populations, who would have less reason and opportunity to turn to non-regulated substances and whose use level could be rigidly controlled. At the same time, the majority of Chinese users were vilified: they undermined the lofty purposes of the colonial state's opium trade and/or brought on themselves problems of addiction (van Dedem, 1881; Meijring, 1974, pp. 73–101; Rush, 1977; Vanvugt, 1985; van Ours, 1995; Chandra, 2002; van Luijk & van Ours, 2007).

Drug Regulations in Suriname

On the other side of the globe, in the Dutch West Indies, by 1900 drug policies were aimed at regulating consumption of opium and cannabis by immigrant indentured labourers from China, British India, and the Dutch East Indies. These groups had brought their drug habits to Suriname. From 1872 onwards, excises had to be paid on the import of their drugs. The result was a thriving smuggling trade, especially with neighbouring British Guyana where use and sale of these drugs had already been regulated in 1861, stimulating illegal production and distribution. In the 1890s, Dutch employers in Suriname started to complain about reduced work performances because of drug use among immigrant labourers, while the colonial mental hospital received an increasing number of patients with reported cannabis-induced insanity. In 1896, a new regulation stipulated that opium and cannabis could only be sold on medical prescription. Smuggling increased, and to control the trade in 1908 a new regulation prohibited the import and sale of cannabis and opium except for licensed wholesalers, apothecaries, and retail dealers; possession of more than small amounts of opium for personal use was criminalized. In 1910, all opium dens were closed, and in 1918, a complete prohibition of the sale and use of cannabis (except by apothecaries and on medical prescription) and opium was enacted in the colony, anticipating the *Opiumwet* (Opium Act) in the Netherlands of the next year (Snelders, 2021, p. 27). The second Dutch Opium Act of 1928, making use, production, and trade illegal, was enacted in Suriname as well. However, illegal use of cannabis and opium continued to exist (Pronk, 1962). After the Second World War and especially after the independence of Suriname in 1975, trade and use of illegal drugs by Surinamese migrants in the Netherlands was prominent in images and realities of the illegal drug market, while the independent state of Suriname would become a major drug trafficker of its own after the military coup of 1980 (Snelders, 2021).

Epilogue: Drug Use and the Colonial Cultural Archive

After the Second World War, available treatments turned leprosy in a so-called neglected tropical disease: not extinct but also not the major public health problem of colonial times. Drug use, however, has only grown despite an increasing war on drugs since the First World War. The results of the colonial framing of drug use are still very much around us, and present in what has been called the 'colonial cultural archive', the narratives and histories in our culture. Let me give an example from the city I grew up in. In the 1980s, the city centre of Amsterdam had become almost a 'no-go' area for respectable and law-abiding citizens. The heroin epidemic and the economic crisis had reshaped the Nieuwmarkt area to the east of Dam square and the Royal Palace. In a neighbourhood characterized by boarded up and squatted houses, heroin junkies, prostitutes, dealers, and pickpockets seemed to prowl the streets. A significant number of them were not white: Surinamese, Moluccans (whose parents had fled Indonesia after independence), and Chinese had a conspicuous presence, adding to the general atmosphere of danger. And in fact, walking the Damstraat from the palace to the infamous 'pill bridge' where clients gathered to purchase illicit drugs entailed the risk, especially if one was a tourist, of losing one's valuables on the way.

This is one of the contexts in which we should evaluate the cartoon that Dutch underground comic artist Peter Pontiac made in the mid-1980s, part of a series that chronicled the life and adventures of two white punk rock rebels: a drawing he entitled '*Heavy shirtje!*' (Heavy shirt!). (We can date the drawing by the presence of a poster hanging on a street wall agitating against the attempts of the city government to bring the Olympic Games to Amsterdam,)

In the cartoon, we see punk couple Gigi and Gaga walking together in the Damstraat (in the background Dam square is clearly recognizable). The male, Gaga, is wearing a newly bought t-shirt, probably purchased in the punk t-shirt store that was located at the beginning of the street. Gaga is clearly content with the shirt that would definitely shock the bourgeoisie. A South African apartheid regime leader is pictured screaming while surrounded by burning car tyres, with under the picture the legend 'Alles sal reg kome!' ('Everything will be alright'), an infamous saying of a white South African politician at the time. Gigi and Gaga are both white. There are also two black people in the drawing. One is a tourist in a suit, taking a photograph of the couple. The other is a scary street robber and pickpocket who has just taken Gaga's wallet without being observed.

If we are looking for drugs in our colonial cultural archive, we hardly need to look further than this underground comic drawing. There are references to both the racist apartheid regime of South Africa that was constructed by the descendants of Dutch colonizers, and to the presence and problematic behaviour of some black Surinamese migrants that had come to the Netherlands from one of its former colonies. Bringing this together in one drawing brought accusations of racism by activists from the left-wing squatter and action movement (to whom Pontiac himself belonged) on the head of the artist. While he himself had meant the drawing as ironic ('what would the anti-racist Gaga say when his wallet was stolen by a black?') he undertook as he said 'self-censorship' and did not publish the drawing at the time (Pontiac, n.d.).

Pontiac's drawing can be read from multiple perspectives: our own, in hindsight, but also the perspectives of many actors at the time who have left their views or whose memories can still be collected. For the contemporary observer at the time, one of the contexts the drawing referred to was the drug problem and its relationship with Surinamese migrants. I believe both that Pontiac was sincerely anti-racist and that the juxtaposition of images in his drawing show how certain associations of a racist character with drug use and abuse, with drug use as a social problem in Gussow's sense, had pervaded our social unconscious. This as part of a much longer history in which Dutch colonialism had shaped colonial societies and our perceptions of and dealings with what became illegal drugs, and in which colonial society in its turn had influenced Dutch culture.

Conclusion

The framing of drug abuse and leprosy as socially threatening conditions was not purely a matter of localized responses. Imperialism meant the need for more control of geographical territories as well as of global migration movements. A global racial framing, connecting the conditions with people of colour and 'inferior races', was a central part of the process of understanding both conditions but also of stigmatizing drug users and leprosy sufferers. By 1900, in the Dutch imagination the opium user had become the languid and passive degenerate consumer in an opium shed, quite akin to the passive leprosy sufferer. Both images were often of people of colour; and consequently, both had to be disciplined in the views of colonial medicine and colonial rulers.

References

Barrell, J. (1991). *The Infection of Thomas De Quincey: A Psychopathology of Imperialism*. Yale University Press.

Bashford, A. (2004). *Imperial Hygiene: A Critical History of Colonialism, Nationalism and Public Health* (p. 2004). Palgrave Macmillan.

Becker, H. (1963). *Outsiders: Studies in the Sociology of Deviance*. The Free Press of Glencoe.

Bewell, A. (1996). *Romanticism and Colonial Disease*. The Johns Hopkins University Press, 1999.

Booth, M. (1996). *Opium: A History*. St. Martin's Griffin.

Chandra, S. (2002). The role of government policy in increasing drug use: Java, 1875–1914. *Journal of Economic History, 62*, 1116–1121.

de Jong, J. J. P. (1998). *De waaier van het fortuin. Van handelscompagnie tot koloniaal imperium*. Sdu.

de Quincey, T. (1821). *Confessions of an Opium Eater*. Penguin (repr. 1986).

Dedem, W. K. baron van. (1881). *Eene bijdrage tot de studie der opiumquaestie op Java. De officiëele literatuur*. J.H. de Bussy.

Dikötter, F. L., Laamann, L., & Xun, Z. (2016). *Narcotic Culture: A History of Drugs in China* (Rev. ed.). Hurtst & Company, et al.

Douwes, A. (2020). *Visiting Pharmacies: An Exploratory Study of Apothecary Shop as Public Spaces in Amsterdam, c. 1600–1850*. MSc-Thesis History and Philosophy of Science, Utrecht University.

Edmond, R. (2006). *Leprosy and Empire: A Medical and Cultural History*. Cambridge University Press.

Foucault, M. (1995). Geschiedenis van de waanzin in de zeventiende en achttiende eeuw. Dutch translation of *Histoire de la folie* (1961). Boom.

Gaastra, F. S. (2002). *De geschiedenis van de VOC*. Walburg Pers.
Gussow, Z. (1989). *Leprosy, Racism, and Public Health. Social Policy in Chronic Disease Control*. Westview Press.
Hoefte, R. (2014). *In Place of Slavery: A Social History of British-Indian and Javanese Laborers in Suriname*. Palgrave Macmillan.
Lampe, P. H. J., & Simons, C. (1929). Lepra in Suriname. *Nederlandsch Tijdschrift voor Geneeskunde, 73*, 4903–4915.
Leung, K. C. (2008) Ki Che Leung, *Leprosy in China: A History*. Colombia University Press, 2008.
Madancy, J. A. (2004). *Troublesome Legacy of Commissioner Lin: The Opium Trade and Opium Suppression in Fujian Province, 1820s to 1920s*. Harvard University Press.
Meijring, K. H. (1974). *Recht en verdovende middelen*. VUGA-Boekerij.
Padwa, H. (2012). *Socila Poison: The Culture and Politics of Opiate Control in Britain and France, 1821–1926*. The Johns Hopkins University Press.
Pontiac, P. (n.d.). *The Pontiac Review*, Vol. 1, *Gaga en de hippieteit!*. Het Raadsel.
Pronk, J. B. (1962). *Verkenningen op het gebied van de criminaliteit in Suriname*. Martinus Nijhoff.
Quinian, S. (1996). Colonial Encounters: Colonial Bodies, Hygiene and Abolitionist Policies in Eighteenth-Century France. *History Workshop Journal, 42*, 107–126.
Rosenberg, C. E. (1989). Disease in History: Frames and Framers. *The Milbank Quarterly, 67*(1), 1–15.
Rush, J. R. (1977). *Opium Farms in Nineteenth Century Java: Institutional Continuity and Change in a Colonial Society*. Ph.D. Thesis, Yale University.
Schiebinger, L. (1989/1990). The Anatomy of Difference: Race and Sex in Eighteenth-Century Science. *Eighteenth-Century Studies, 23*, 387–405.
Schilling, G. W. (1771). *Verhandeling over de melaatsheid*. J. C. ten Bosch.
Snelders, S. (2013). Leprosy and Slavery in Suriname: Godfried Schilling and the Framing of a Racial Pathology in the Eighteenth Century. *Social History of Medicine, 26*(3), 432–450.
Snelders, S. (2017). *Leprosy and Colonialism: Suriname Under Dutch Rule, 1750–1950*. Manchester University Press.
Snelders, S. (2021). *Drug Smuggler Nation: Narcotics and the Netherlands 1920–1995*. Manchester University Press.
Snelders, S. (in preparation). Inleiding. *Jaarboek De Achttiende Eeuw: Drank & drugs in de achttiende eeuw*.
Snelders, S., van Bergen, L., & Huisman, F. Leprosy and the Colonial Gaze. (2021). Comparing the Dutch West and East Indies, 1750–1950. *Social History of Medicine, 34*(2), 611–634.
Szasz, T. (1985). *Ceremonial Chemistry: The Ritual Persecution of Drugs, Addicts, and Pushers*. Learning Publications.

van Bergen, L. (2018). *Uncertainty, Anxiety, Frugality: Dealing with Leprosy in the Dutch East Indies, 1816–1942.* NUS Press.

van den Doel, H. W. (1996). *Het Rijk van Insulinde. Opkomst en ondergang van een Nederlandse kolonie.* Prometheus.

van Doorn, J. A. A. (1994). *De laatste eeuw van Indië.* Bert Bakker.

van Driel, B. M. (1922). Over behandeling en bestrijding van lepra. *Geneeskundig Tijdschrift voor Nederlandsch-Indië, 62,* 149–163.

van Hasselaar, A. (1835). *Beschrijving der in de kolonie* Suriname *voorkomende elephantiasis en lepra (melaatschheid).* S. de Greber.

van Lier, R. (1971). *Frontier Society: A Social Analysis of the History of Suriname.* Martinus Nijhoff.

van Luijk, E. V. V., & van Ours, J. C. (2007). The Effect of Government Policy on Drug Use: Java, 1875–1904. *Journal of Economic History, 61,* 1–18.

van Ours, J. C. (1995). The Price Elasticity of Hard Drugs. *Journal of Political Economy, 103,* 261–279.

Vanvugt, E. (1985). *Wettig opium. 350 jaar Nederlandse opiumhandel in de Indische archipel.* In de Knipscheer.

Wright, H. P. (1889). *Leprosy, an Imperial Danger.* Churchill.

PART II

Tropical Medical Institutions and Imperial, Commercial, and Political Expansion

CHAPTER 4

The Business of Tropical Medicine: Connections Between Anti-malarial Campaigns in Sierra Leone, 1899–1901, and Jamaica, 1908

Juanita De Barros and Deborah J. Neill

In 1902, an editorial in the Jamaican newspaper, the *Daily Gleaner*, commented on the work of Alfred Jones, the ambitious Welshman who ran the Elder Dempster Shipping Line and was Chairman of the Board of the Liverpool School of Tropical Medicine (LSTM). Jones, the editors noted, had shown strong support for anti-malaria activities in West Africa where 'practical wonders [were] being effected' including the falling of malaria infection rates in Sierra Leone. In that colony, 'the whole of the citizens'

J. De Barros
Department of History, McMaster University, Hamilton, ON, Canada
e-mail: debarr@mcmaster.ca

D. J. Neill (✉)
Department of History, York University, Toronto, ON, Canada
e-mail: dneill@yorku.ca

M. Capocci, D. Cozzoli (eds.), *Empire, Nation-building, and the Age of Tropical Medicine, 1885–1960*, Medicine and Biomedical Sciences in Modern History,
https://doi.org/10.1007/978-3-031-38805-7_4

had become 'interested in the campaign.' Desperate for Jamaica to implement its own anti-malarial initiatives, the editors proclaimed that 'we need the influence of Sir Alfred Jones in Jamaica' (*The Daily Gleaner* (hereafter *DG*), January 28, 1902, 4). They were not disappointed. Six years later, Jones facilitated a major LSTM medical expedition to Jamaica, leading the colony's former governor, Henry Blake, to enthusiastically endorse the 'great work' that Jones was doing in 'promoting the welfare of the colonies' (*DG*, December 2, 1908). The expedition aimed to combat tropical diseases, particularly malaria, and was headed by the former public health officer of Sierra Leone, William Prout.

Although medical researchers were determined to eradicate several dangerous diseases associated with the 'tropics'—including sleeping sickness, yellow fever, and ancylostomiasis (hookworm disease)—they and others viewed malaria as particularly pernicious. The surgeon general for British Guiana and later Trinidad, K.S. Wise, maintained that it was the most 'incapacitating disease to which man [was] liable' and that its effects were both mental and physical (Wilkinson & Hardy, 2001, p. 51; Wise, 1919, pp. 5–7, 13–15). Tropical medicine expert Ronald Ross, who began to seriously investigate malaria while in the Indian Medical Service, believed it was the most important disease to conquer as it was 'the great sapper of civilisations [and] the destroyer of imperial races' (Arnold, 1999, p. 127). In 1898, Ross made a crucial breakthrough when he identified the role of the *Anopheles* mosquito in carrying the microorganism that caused the disease. He soon joined the LSTM and the school began to organize mosquito eradication campaigns in many diverse sites, including India, the Middle East, West Africa, and the Caribbean (Nye & Gibson, 1997).

Ross hoped that his methods of vector control could be universally applied, reflecting a similar ethos to the 'Pastorians' who believed their microbial containment methods could be successful 'independent of race, environment, or society' (Velmet, 2020, p. 7). This strategy diverted attention from broader (and more costly) public health solutions and marked a shift away from considerations of environment and the need for 'social uplift' for affected populations (Packard, 2007, pp. 115–126; Webb, 2009). K.T. Silva's study of rural Sri Lanka shows that emerging scientific ideas, including those of Ross, 'became identified with the interests of the colonial rulers and the planters' and contributed to this shift (Silva, 1997, p. 203). These scientific and capitalist 'empire builders' not only shaped the priorities of malaria eradication campaigns but also facilitated the transfer of ideas from one region to another, however different those regions might be (Lambert & Lester, 2006; Neill, 2012, Velmet, 2020). Andindeed, economic factors, such as trade, had long influenced medical developments (Harrison, 2010).

The kind of relationships between researchers, capitalists, and officials that these researchers have identified existed elsewhere. Focusing on Jamaica and Freetown, Sierra Leone, our chapter explores the influence of these interactions on public health policies, particularly regarding malaria. We examine which anti-malarial strategies were deployed in both places, and how approaches used in one can help to explain measures taken in the other. The two sites were quite different. Freetown was a densely populated urban centre, whereas Jamaica was a more sparsely inhabited plantation colony. The elites of both societies, moreover, had different histories and ambitions. Despite these differences, the involvement of Ronald Ross, William Prout, and Alfred Jones provides a unifying factor. All three came from outside malaria zones with the ambition of conquering tropical diseases (Jones, 2013, pp. 95–100; Macqueen, 2005, p. 217). While each had an influential role, we also demonstrate that the actual implementation of measures in Sierra Leone and Jamaica was far from a top-down imposition of one-size-fits-all malaria policies. Common ideas motivated some approaches to combating malaria in each place, but local political, economic, and scientific interests played an important and complicating role. While the same men had influence, input from multiple factions was vital to the cross-fertilization of ideas and the implementation of specific policies from one imperial space to another.

The Liverpool School of Tropical Medicine as Hub

Both the Liverpool School of Tropical Medicine and the London School of Tropical Medicine were established at the end of the nineteenth century to fight tropical diseases, part of the push of the British Colonial Office for more 'development' in the colonies (Power, 1999; Wilkinson & Hardy, 2001; Worboys, 1988). The founding of these institutions demonstrates that strategic imperial goals, including the extension of trade and settlement, underlay British anti-malaria initiatives across the tropical world, something that contemporaries recognized. Following the return of the LSTM expedition from Jamaica in 1909, the Liverpool Journal of Commerce noted the 'extraordinary work' undertaken by Jones and 'his scientific colleagues' at the LSTM in 'discovering the means of making tropical countries healthy for Europeans and Americans' (*DG*, July 3, 1909, 17). Complaints by British merchants about health problems in West Africa, increasing acceptance of the merits of tropical medicine research, and the appointment of the renowned physician and researcher Patrick Manson as medical advisor to the Colonial Office contributed to the growing institutionalization of tropical medicine (Dumett, 1968; Worboys, 1988). As John Farley has observed, tropical medicine was part of 'empire building and empire maintenance' (Farley, 2003, p. 4).

Alfred Jones, who first proposed the creation of the LSTM, had banana plantations in the Canary Islands, hotels, warehouses, and coaling companies across Western Africa and in Jamaica, and a controlling interest in the British Bank of West Africa. He saw in the school the possibility of realizing his ambition to gain the ear of the Colonial Office for money-making ventures and to conquer a disease in the name of commerce. 'Money spent in our School of Tropical Medicine is an investment,' he once said, 'and we expect dividends from it' (quoted in Lyons, 1992, p. 70). In a 1909 speech, he also declared that 'we must never miss an opportunity of combining science with commerce' (*DG*, July 2, 1909, p. 1). When the school initially struggled to gain government support, Jones raised start-up funds from Liverpool's university, hospital, and business community (primarily via the Chamber of Commerce). He also turned to corporate donors in other parts of Britain, as well as Germany, France, and West Africa, offering named buildings, tributes, and medals in exchange for money that funded infrastructure and endowed professorships (including Ronald Ross's appointment in 1899).[1] This reliance on the business community, as Michael Worboys has shown, led to the scientists promoting their work as a 'consultancy' and their expeditions as 'entrepreneurial activities' to control malaria across 'plantations, factories and trade' (Worboys, 1988 pp. 28, 27, 24).

The earliest LSTM expeditions were sent to the British-controlled city of Freetown, Sierra Leone, while a subsequent one went to Jamaica. In an interview in 1899, Jones emphasized the importance of the LSTM's work in mitigating the negative health effects of west Africa's climate; he noted

[1] Jones was the largest donor, but others include The British and African Steam Navigation Co., the African Direct Telegraph Co., Hatton & Cookson, John Holt, Hamburg-based shipping firm Woermanns, and many other British, French, and African firms. See *The First Report of the Liverpool School of Tropical Diseases*, Appendix 2, 'List of Subscriptions and donations, 1899,' Bodleian Library, Oxford (hereafter Bod.Lib.), MSS Afr. S 1525 Box 11/6.

the possibility that England might reap 'millions upon millions of gold' if solutions to tropical diseases were found ('England's Mistakes in West Africa: Mr. A.L. Jones: Views,' *Sierra Leone Weekly News* (hereafter *SLWN*), 14 October 1899). His ambitions were shared by some of the LSTM physicians, including Dr. Fielding-Ould, who stated that once malaria was conquered, 'Liverpool ought to get the largest share' of west Africa's bounty ('The Malaria Expedition. Interview with Dr. Fielding-Ould,' Liverpool Daily Post, 1 January 1900, reprinted in the *SLWN*, 27 January 1900). The first expedition, led by Ross and financially supported by Jones and other corporate donors, was tied to these capitalist dreams, a pattern that would be reflected in the later expedition to Jamaica.[2]

Freetown, 1899–1901

Over the course of the nineteenth century, the city of Freetown had developed into a bustling port with a diverse population that included the Temne and Bullom/Sherbro peoples, as well as a variety of settlers frequently referred to as Creole/Krio and including the 'Nova Scotians,' 'Maroons,' and Africans liberated from slave ships (Cole, 2013).[3] Its strategic location drew merchants and workers from along the coast and hinterland. Rapid growth meant building construction outpaced city services, and in many places the city lacked clean piped water, organized sewage disposal, and paved roads. Freetown's dismal infrastructure had contributed to a sinister reputation for high death rates; as the historian Leo Spitzer notes, the city was deemed so unhealthy 'that insurance companies refused to insure the lives of Europeans going there' (Spitzer, 1968, p. 51).

William Prout had been appointed the Chief Medical Officer in 1895. An 1884 University of Edinburgh graduate, Prout had previously served as a medical officer in Mauritius, assistant colonial surgeon in the Gold Coast, and head of the medical department in Gambia ('Obituary: Sir

[2] Aside from Jones's contributions, funding came primarily from merchants and the British Medical Association. Ross later persuaded Glasgow thread magnate James Coats to donate a further £2000. For more on donors, see Ronald Ross (1901), pp. 3–4.

[3] The 'Nova Scotians' were free and liberated Africans returning to West Africa; the 'Maroons' came from Jamaica via Nova Scotia. Among the many other groups were Igbo, Ope, and Fante. For further insights on the peoples of Sierra Leone, see Jalloh (2018) and Bangura (2017).

William Prout,' *The Times*, 20 November 1939, 8). His medical reports make for grim reading, as he recorded high death rates from a variety of diseases, identifying smallpox, digestive and bronchial problems, 'fevers,' and infant mortality as the most serious challenges. The thorough 1897–1898 report noted the difficulty of the intertwining factors that led to illness, including poor nutrition, poor quality housing, and the absence of a city-wide sewage system (Prout, 'Sanitary Report on the City of Freetown 1897 & 1898,' Liverpool School of Tropical Medicine (hereafter LSTM), ROSS/67/01). Prout, therefore, had advocated repeatedly for infrastructure investments including improved surface drainage via culverts and paving, improved cesspits and latrines, and the expansion of indoor plumbing. He also expressed the need for better building codes and called for attention to infant and child mortality. But he had seen very limited success. 'Of all the suggestions which I have made in my last report,' he lamented in 1898, 'I am unable to state that a single one has been carried out' (Prout, 'Sanitary Report on the City of Freetown for the Year 1898,' LSTM ROSS/67/01, p. 27).

Prout's focus on sanitation and public health had the support of several prominent Freetonians, including Senior Assistant Surgeon Dr. William Awooner-Renner who was president of the quarantine board and the likely author of several of Prout's annual reports, although he did not get the credit (Patton, 1996, pp. 135–136). He was also supported by J.A. Fitz-John, the editor of the *Sierra Leone Times*. Fitz-John argued that the water supply question was 'so vital and so urgent that all other considerations should have been made to stand until it be undertaken' ('The Water Supply,' *Sierra Leone Times* (hereafter *SLT*), 18 March 1899). But Freetown's diverse population was divided, with one newspaper article even claiming that hinterland immigrants were 'sworn foes … to even the most elementary laws of cleanliness and decency' ('The Health of Freetown,' *SLT*, 14 March 1896). Governor Charles King Harman seized on this disunity to dismiss calls for city-wide spending on sewers and piped water. He argued that local people were 'not adapted' to sanitary infrastructure and could not 'use any system thoroughly' (Governor Charles King-Harman, as quoted in the *SLWN*, 30 May 1903, 3). LSTM expedition scientists such as Logan Taylor also expressed this view. He admitted that mosquitos had been seen breeding in poor quality cesspits, but he rejected the installation of new latrines since nightsoil removal was unsightly and claimed that local people 'did not see the necessity of cleanliness' (Letter from M. Logan Taylor to RR, Freetown, 19 December 1901, LSTM ROSS/79/75/112; LT to RR, notes on a report, LSTM

ROSS/113/07/05). It is small wonder, given these views, that little infrastructure development had been done.

The scientific team that first arrived in 1899 sought to use Freetown as a laboratory in which they could prove that it was possible to eradicate the *Anopheles* mosquito and thereby reduce the malaria case load.[4] The first step was to identify the vector. Local newspapers reported with enthusiasm that the team had confirmed an abundance of *Anopheles* throughout the city. The next step was to destroy them, and Ross employed teams of street cleaners known as 'mosquito brigades' who went from house to house, emptying water cans, bottles, buckets, and gutters, and using kerosene to 'oil' larger pools of water, thereby inhibiting larvae breeding. The team also distributed nets and prophylactic quinine so that individuals could personally protect themselves. Beyond that, however, they showed little interest in Freetown's public health system. Ross breezily told Jones that he was 'having a look round at the other medical and sanitary matters—drainage, conservancy &c,' but he quickly returned to destroying mosquitos, claiming that 'most of the malarial fever here can be got rid of at almost no cost except of a little energy on the part of local authorities' ('The Malarial Mission,' *SLWN,* 21 October 1899, p. 8). House-to-house work went on for the duration of the first mission and resumed with the arrival of a second team in 1901, which included Taylor.

Unlike in Havana, where William Gorgas's massive eradication campaign had dramatically reduced yellow fever, there was no appreciable results from the constant cleaning work. The incessant rains, dense vegetation, and the population density made things challenging. Taylor claimed that their efforts had been significantly hindered by Prout and the Freetown Municipal Council. He complained that Prout 'pretended to help us but really was an active opponent behind our backs' (Logan Taylor to RR, 24 November 1905, LSTM ROSS/113/07/04; Notes from Logan Taylor to RR, 1905, LSTM ROSS/113/07/05). Ross was equally accusing, arguing that Prout had 'thwarted all my work at Freetown' and telling Joseph Chamberlain that he had shown 'an extremely moderate interest in anti-malaria work, and none at all in ours' (Note on files about Prout by RR, LSTM ROSS/File 32; RR to Colonial Secretary Joseph

[4] The 1899 group included Ross, A.E. Annett, E.E. Austen, and the Belgian-sponsored G. van Neck. See Austen (1899). The 1901–2 expedition included Ross, Taylor, C.W. Daniels, and Anderson McKendrick. See Logan Taylor (1902), p. 7.

Chamberlain, 9 July 1903, ROSS/77/26). Certainly, Prout disagreed with some of the team's methods. In a 1903 letter to the *British Medical Journal*, Prout argued that the constant rains rendered street brigades ineffective, and that what was needed were improved cesspits and a 'system of permanent surface drainage, which I advocated so long ago as 1896.' Prout acknowledged Ross's role in drawing attention to the malaria problem, and noted the Governor's drainage plans for some districts, but he argued that change could only come with African cooperation, and that visiting experts should not be 'belittling the work of others in the same direction.'[5]

Prout's allies in the locally run Municipal Council and in the press also expressed concerns. Freetonians worried that the malaria brigades were a poor substitute for bigger investments in the sanitary system. The *Saint James Gazette* reminded readers that 'the only method of getting rid of the mosquitos is the same as that of getting rid of malaria, namely, to drain the marshes' ('Common Sense about Mosquitos,' *Saint James Gazette*, as reprinted in *SLT*, 28 October 1899). The *Sierra Leone Weekly News* worried that 'so long as the roads and streets remain level and flat and the drains such as they are, we can never hope for any change for the better.' Almost word for word, the same point was made again in 1904, when the author noted that only in Grassfields had the streets been improved ('The Seasons and their Effects,' *SLWN*, 8 July 1899; 'The Seasons and their Effects,' *SLWN*, 3 September 1904, 4). These critiques took aim at a short-term approach they saw as aligned more with proving a scientific theory than addressing Freetown's overall health needs. Ross did hint at the difficulties in a letter to Jones in September of 1901. While he still insisted that 'the game is well worth the candle' and 'the results already well marked,' he acknowledged that 'torrents of rain' had caused problems (Quoted in 'The Malaria Expedition,' *SLWN*, 28 September 1901, 4). Taylor more frankly admitted that they were not succeeding, noting that '[U]nless the Government and the school will keep on the work, the money the school has spent on it will be almost thrown away' (Quoted in 'The Extermination of the Mosquito in Sierra Leone,' *British*

[5]William Prout, 'The extermination of Mosquitos in Sierra Leone.' *British Medical Journal*, vol. 1, no. 2214, 6 June 1903, p. 1349. Prout was responding with objections to an update provided by Logan Taylor, *British Medical Journal*, vol. 1, no. 2194, 17 January 1903, pp. 156–157. Taylor wrote a furious rebuttal, but the journal did not publish it, claiming length restrictions. See D. Williams to RR, 30 June 1903, LSTM ROSS/74/06.

Medical Journal, vol. 1, no. 2194, 17 January 1903, p. 156). Scientific quarrels may have hindered the cleaning work, but in the meantime another plan was taking shape, originating in the governor's office. King-Harman's reluctance to invest in paving and sewer upgrades was deeply tied to one of his 'pet projects,' which was segregation ('The Governor of Sierra Leone and the Manchester Chamber of Commerce (African Section),' *SLWN*, 30 May 1903, 3). Already in 1901 he had commended Ross for 'the good work you are doing in staving off the commission with their sanitary engineer,' calling it a 'useless burden' and insisting that Sierra Leone 'is too poor to pay for such luxuries.' Instead, he told Ross, he had convinced the Colonial Office to fund waterworks and a railway leading into the hills above Freetown, where he planned for 'the erection of suitable residences for Europeans' (Governor Charles King-Harman to RR, 20 October 1901, LSTM ROSS/79/65/2/3).[6]

The governor freely acknowledged that he was pursuing segregation in the face of 'a large amount of opposition in Freetown' (Frenkel & Western, 1988, p. 211; 'The Governor of Sierra Leone and the Manchester Chamber of Commerce (African Section),' *SLWN*, 30 May 1903, 3). Obtaining scientific backing therefore helped him justify his decision. Ross had always supported segregation as a possible measure—in fact he was struck by 'how much Europeans in Africa suffered from not being segregated from the natives as they are in India.'[7] But the team had not included segregation as a primary strategy in its initial reports. This changed as the brigades failed to achieve their objectives and Liverpool colleagues like Fielding-Ould publicly supported the re-location of Europeans to 'the healthy places' ('The Malaria Expedition. Interview with Dr. Fielding-Ould,' *Liverpool Daily Post*, 1 January 1900, reprinted in the *SLWN*, 27 January 1900, 2). For his part, Jones indicated that he would move his own employees there, although he was one of only a handful of business

[6] The commission was the Malaria Commission of the Royal Society, of which Ross was a member.

[7] Ronald Ross (1910), p. 45. Ross saw segregation as more of a temporary solution. His listed priorities in 1910 were first to drain marshes and rid urban centres of standing water and distribute quinine and use nets. It was the Colonial Office and the Governor of Sierra Leone, who used the Royal Society malaria commission recommendations, who prioritized the physical separation of Europeans from Sierra Leonians. See Frenkel and Western (1988), pp. 215–216.

leaders to do so. In fact, merchants like John Holt and Richard Blaize spoke out against segregation in the press.[8]

As a public health measure, segregation was a failure. The first 20 bungalows, with their dedicated water supply, cost £46,000; and for the 90 of the 558 Europeans who moved to the hills, rates of malaria did not significantly decline (Spitzer, 1968, pp. 57–58; Goerg, 1998, pp. 16–18). It also fuelled considerable local resentment. The *Sierra Leone Times* lamented 'what a splendid thing the Government could have made of our streets and public buildings' if the funds had been differently allocated ('One Thing and Another,' *SLT*, 16 January 1904). Another writer noted that 'by draining Freetown scientifically a more lasting and general benefit would be conferred,' whereas this plan would benefit a tiny, transient minority of Europeans. 'Every government,' this letter-writer noted, 'should float or sink with its people and not endeavour to cut itself adrift from them' (Anon., 'Sierra Leone and the Mountain Railway,' *SLWN* 4 October 1902, 2).

This divisive period led Prout to shift his allegiances. From a champion of public health infrastructure and roadworks, he became, by the time he joined the Jamaica expedition in 1908, an advocate of both segregation and LSTM anti-malarial work. As early as 1902, while he acknowledged that 'the improvement of the health of the natives is of equal importance' to that of the Europeans, he called segregation 'the most important step in the direction of health preservation which has yet been taken' for Europeans ('Annual Report of the Medical Department of Sierra Leone 1902,' *SLWN*, 1902, 5). By 1908, the *Sierra Leone Guardian* was complaining that he was 'strongly recommending the Bungalows and Hill Stations in order to keep the white population apart and together,' and that he 'represents Freetown as full of Malaria and residence therein, being surrounded by houses so infected, is far from anything like safety' ('The System of Segregation,' *Sierra Leone Guardian*, 10 April 1908, 4). Aligning himself with the policies of the Colonial Office and the governor

[8]'The Malaria Expedition,' *SLWN*, 28 September 1901, 4; 'The Governor of Sierra Leone,' 2. Holt hoped Ross would succeed, which would perhaps 'avoid the necessity of the segregation of Europeans, a thing he did not agree with.' Yoruba merchant Blaize said that 'Africa could not be governed on the lines which were practicable in India,' and that segregation was a 'theoreticum absurdum.' See 'Liverpool African Trade Section: Sanitation and Trade, Cable Rates and Customs Tariffs,' reprinted in the *Lagos Weekly Record* 22 June 1901; 'The Liverpool Courier and an African Merchant,' *SLWN*, 28 July 1900, 5; 'The Liverpool School of Medicine and Segregation,' *SLWN*, 28 September 1901, 5.

was not his only goal; he also increasingly echoed the priorities of the LSTM scientists by more expressly emphasizing malaria as the primary problem of the tropics. Even in his 1902 medical report, he argued that 'it was extremely probable' that malaria was the major killer of young children ('Annual Report of the Medical Department of Sierra Leone 1902,' *SLWN* 30 April 1904, 4). He would soon link malaria to a decline in productivity and profits, maintaining that solving malaria would boost economic growth.

Prout's shift points to the medical community's broader transition away from 'social uplift' ideas to narrow biomedical solutions. It also demonstrates how the combined pressures applied by political, economic, and scientific stakeholders could alter a scientists' views. Beginning in 1905, Prout had become well aware of the need to get along with Ross. That year, Alfred Jones offered him an LSTM lectureship and he quickly resigned his post in Freetown ('Copy of Application from Dr. W.T. Prout to be attached to the Staff of the Liverpool School of Tropical Medicine,' 17 May 1907, University of Liverpool (hereafter UL) P4A/514). But Jones had not consulted the School's board, and Ross was furious, telling Jones that he was 'not responsible for your independent engagements, and I am not in the smallest degree concerned with his position or prospects' (RR to Sir Alfred Jones, 9 April 1908, LSTM ROSS/113/07/36). Ross also raged to the School's Honorary Secretary that Prout's 'skeptical and indeed inhibitory attitude' had been harmful to his mission. Prout found himself left in limbo (RR to Milne, 27 February 1907, LSTM ROSS/113/07/09). In May of 1908, Prout again begged Jones to intervene, noting that while he didn't want to 'embarrass you,' he was incurring a 'very serious financial loss' (William Prout to Sir Alfred Jones, 20 May 1908, UL P4A/514). Jones then supported Prout's appointment to the Jamaica expedition, presumably as a means of further postponing the promised lectureship. When Jones held a send-off lunch, Prout gave a speech that emphasized his loyalty to the school and to Ross, receiving applause when he stated that 'if any improvement had been affected [in malaria rates] it had been mainly due to the work of the Liverpool School and the epoch-making researches of Professor Ross' ('Sanitation in Sierra Leone,' *Liverpool Courier* clipping in ROSS13/160).

The story of anti-malaria measures in Freetown points to important intersections between scientific, political, and capitalist priorities; racial bias; elite influence; and the intersection of institutional and personal goals. While there were some road and water improvements, the work was slow and fragmented. Freetonians pushed for city-wide sanitation measures, but the governor and the Colonial Office instead pursued their idea of a whites-only suburb, and the scientists, despite internal disagreements, ultimately endorsed the governor's priority that proved to be a failure as a public health measure.

The men involved in crafting Freetown's anti-malaria strategies were keen to bring their ideas and experience to other parts of the British empire. This imperial context is important for understanding how business, politics, and science meshed in many places. But as the next section demonstrates, while building on what they had learned elsewhere, Prout and the Liverpool team would also be strongly influenced by local politics and the context of malaria within a rural setting and a plantation-heavy economy.

Jamaica

Many of the patterns seen in Sierra Leone were apparent in Jamaica. In both places, newspapers weighed in on the need for public health reforms, often channelling the views of local economic and political figures, including physicians. Worries about the economic impact of diseases such as malaria animated them. These colonial contexts were important. Despite the imperial connections that Prout and Jones enjoyed and their enthusiastic efforts to realize their own career (and financial) goals, colonial realities significantly shaped the results of their undertakings.

In Jamaica, anxiety about the colony's economic health underlay British and Jamaican responses to Jones's activities in West Africa and his expressions of interest in the Caribbean colony and the malaria expedition he helped underwrite. Jones's involvement in the tourist and banana industries convinced imperial figures such as the colony's governor, A.W.L. Hemming, as well as many Jamaicans that the future was bright indeed. He was Jamaica's 'saviour' as Frank Taylor points out his work on the Jamaican tourist industry (Taylor, 1993, p. 85).

Jamaicans also hoped that Jones's interest in solving some of the colony's pressing medical problems would bear fruit. Imperial and colonial officials and members of Jamaica's political and economic elites eagerly anticipated that the Liverpool School's malaria expedition to Jamaica in 1908–1909 would reveal the secret to defeating this economy-sapping disease. Both initiatives were ultimately less successful than Jamaicans had hoped but studying the ways in which they intersected and informed each other reveals the extent to which colonial and imperial goals were mutually constituted. The 1908–1909 Liverpool School expedition to Jamaica reflected the imperial connections and concerns addressed in the first part of this chapter, especially Prout's own medical experience and his conclusions about the causes of and solutions to malaria. But the economic and social reality of Jamaica and the agendas of individual Jamaicans also helped shape this imperial initiative.

In Jamaica, as in Freetown, economic factors helped shape medical priorities. In this British colony in the Caribbean, the economic crisis of the post-enslavement period was especially significant. By the late nineteenth century, officials in both Jamaica and Britain believed that bananas and tourists could save Jamaica from the economic impact of the sugar industry's decline. The end of slavery, imperial trade decisions, and changes in the world sugar market helped produce a crisis in the sugar industry in much of the British Caribbean including Jamaica in the second quarter of the nineteenth century (De Barros, 2014, pp. 43–44). An 1886 British government commission of inquiry suggested that colonies such as Jamaica should look to other areas of agricultural production, such as the fruit industry, and it called on the British government to temporarily subsidize the transportation of fruit between the Caribbean and British and North American markets (West India Royal Commission, 'Part III-Concluding Observations and Summary,' *Report of the West India Royal Commission, with Subsidiary Report by D. Morris and Statistical Tables* [C.-8655] 1897 (House of Commons Parliamentary Papers Online, 2006), 60, 64, 70 [afterwards *Report of WIRC*]; Joseph Chamberlain, 'Commission,' in *Report of WIRC*, iii; Holt, 1992, p. 332). For Jamaica, the banana industry offered the most potential, a development that brought Alfred Jones and his shipping company into the picture. As we will see later in this chapter, although Jones's enterprises were not as successful as he and others had hoped, these commercial activities had a real impact on public health policy, notably that dealing with malaria.

British imperial concerns about U.S. influence helped position Jones as Jamaica's 'saviour,' temporarily at least. Initially produced by Afro-Jamaican peasants, bananas became more significant in the colony from the late nineteenth century, and by the early twentieth century, as an export, it had surpassed sugar. It attracted increasing attention from large interests, such as the United Fruit Company, a U.S. enterprise (San Miguel, 2011; Holt, pp. 124, 347–365. And see 'Jamaica,' in Report of WIRC, 59–62). The British government was keen to ensure that Jamaica's expanding banana industry did not make the colony overly dependent on the United States, and it wanted to encourage tighter links between Jamaica and Britain (Taylor, 1993, p. 84). Secretary of State Joseph Chamberlain turned to Jones whose shipping company was already carrying bananas from the Canary Islands to Britain, a decision that reflected Chamberlain's belief in the role that wealthy capitalists could play (Holt, 1992, p. 354; Will, 1970, pp. 234–235). In 1900, the British government and Jones signed a contract setting out this arrangement which stipulated that the former pay Jones an annual subsidy of £40,000 for 10 years in exchange for his agreement to have his ships regularly carry bananas and passengers between Jamaica and Britain (Holt, 1992, p. 354; Greenhill, 1969, p. 120; West Indies. Jamaica, 'Contract between the Crown Agents for the Colonies and Messrs. Elder, Dempster & Company for a Steamship Service between Jamaica and the United Kingdom,' 1900 [Cd 175], *BPP* (House of Commons PPP Online, 2006) [afterwards, 'Elder, Dempster Contract,' *BPP*], 379). Jones also became involved in developing Jamaica's nascent tourist industry. He acquired several hotels that had been built to house visitors to Jamaica during the 1891 International Exhibition and, hopefully, stimulate the tourism industry (Lumsden, 1991, pp. 17–21; Taylor, 1993, p. 56; Will, 1970, pp. 234–235). Jones saw the hotels as offering opportunities that could align with his shipping interests; in 1901, his shipping company had rented and restored both (Lucas to Ellis, 20 July 1900, CO 137/616, Jamaica no. 23980, TNA; Oliver to Chamberlain, 4 August 1900, CO 137/612, confidential, Jamaica no. 28464, TNA; Elder, Dempster & Co. to Under-Secretary of State for the Colonies, 19 July 1900, CO 137/616, Jamaica no. 23417, TNA; Jones to Pearson, 24 April 1905, CO 137/649, Jamaica no. 13688, TNA). In his correspondence with Colonial Office officials, Jones used the language of economics and empire to justify and defend these ventures, noting the sums he spent and the number of tourists that his hotels attracted (Jones to Pearson, 24 April 1905, CO 137/649, Jamaica no. 13688, TNA; Holt, 1992, p. 355). However, the success of both ventures was short-lived. The shipping line's success was stymied by factors including opposition by supporters of a rival shipping firm and the growing power of the United Fruit Company which steadily acquired interest in and then control of the British organization (Greenhill, pp. 121–126; Holt, 1992, pp. 354–355). The hotels were also disappointing as distance and cost discouraged

Britons from travelling to Jamaica as tourists (Taylor, 1988, p. 209; 1993, pp. 85–87).[9]

Jones's support for the LSTM's medical ventures to Sierra Leone and Jamaica earned him praise from politicians and others in Britain and the empire. They doubtless recognized that tackling diseases like malaria was necessary for business success. On the eve of Prout's departure for Jamaica, the mayors of Liverpool and Manchester sent telegrams expressing their hopes for a successful trip as did officials in the British Colonial Office (Jones to Pearson, 24 April 1905, CO 137/649, Jamaica no. 13688, TNA). Imperial officials were aware of the impact of malaria on Jamaica's economy, particularly its banana industry, and commentators pointed to the impact of these medical ventures on the larger empire (Oliver to Chamberlain, 24 September 1900, CO 137/613, Jamaica no. 33249, TNA. On Ogilvie, see *DG*, 21 October 1911, 14; *DG*, 2 December 1908, 9).

The editors and owners of Jamaica's only daily newspaper in this period, the *Daily Gleaner*, were enthusiastic about the trip, regularly publishing columns and editorials praising the anti-malaria work of Jones and Prout and the importance of addressing public health problems more generally (De Barros, 2014, chapter 5). Since the early twentieth century, the newspaper had been commenting regularly on the same kind of public health issues addressed in Freetown's newspapers (*DG*, 18 August 1905, 5). It also played an important role in calling for public health issues, including those dealing with mosquitos and malaria (e.g., see DG, 28 January 1902, 4; DG, 26 January 1906, 5). It also kept Jamaicans informed about Jones, Elder Dempster, and LSTM initiatives well before Prout's arrival. It described their successful public health activities throughout the empire, hoping that something similar would be introduced in Jamaica (*DG*, 12 March 1094, 8. And see, for example, *DG*, 16 November 1906). Once Prout arrived in the colony, the *Gleaner* publicized his work, interviewing Prout and reporting on his findings (*DG*, 21 January 1909, 8; *DG*, 28 January 1909, 3).

This support may well have reflected the specific views of the newspaper's owners, but it also points to the impact of larger class interests. Unlike Freetown, where Africans were prominent in elite media circles,

[9] On Jones's death, in an editorial in *Gleaner*, the editor noted the slow progress in the banana industry something which the writer was sure that Jones himself 'regretted.' *DG*, 15 December 1909, 6.

the *Gleaner* was owned and edited by members of Jamaica's mercantile elite, part of the colony's Portuguese-Jewish population. Herbert George de Lisser edited the newspaper and members of the founding family (the DeCordovas) owned the paper.[10] From the late nineteenth century, members of this class became involved in the banana industry, either as plantation owners or shippers. Although de Lisser and the DeCordovas did not seem to have been directly involved in the banana trade, the *Gleaner* shareholder Charles E. Mercado was (Bryan, 2000, pp. 70–71). Members of this class, though, did not need to be directly involved in the banana industry to support the public health measures needed for its success. Rather, their shared interest in Jamaica's economic development was sufficient. The newspaper editorialized about various public health problems, especially those that could harm the tourist sector, and regularly called for their amelioration (*DG*, 4 November 1904, 8).[11] The newspaper's editors, along with other members of the colony's business and political elites, understood that this disease was an obstacle to the economic success of the banana and the tourist industries. They were not alone is holding these views; the colony's physicians ascribed to them as well and pushed for public health reforms, particularly those addressing the problem of malaria. They too recognized its economic impact (Island Medical Department, [afterwards, IMD] 'Report on the Island Medical Department for the Year ended 31st March 1906,' *Departmental Reports for 1905–1906*, CO 140/234, TNA, 91, 92; IMD, 'Report of the Island Medical Department for the Year ended 31st March 1907,' *Departmental Reports for 1906–1907*, CO 140/234).

William Prout arrived in Jamaica in November 1908, along with Robert Newstead, a fellow researcher at the LSTM. The School (and Jones) supported this expedition financially (including putting them up in Jones's Jamaican hotels), but so did the Jamaican government. It paid for their travelling expenses while the two men were in Jamaica (*DG*, 13 March 1909, 13; Oliver to the Earl of Crewe, 24 December 1908, CO 137/649, Jamaica no. 11278, TNA). Newstead's job in Jamaica was to investigate

[10] Rush (2004), 26, fn. 44, 191; Birbalsingh (1982).

[11] By the following month, observers concluded that sewers or nearby swamps were the source of the odour. *DG*, 6 December 1904, 6.

cattle ticks whereas Prout was to study malaria.[12] Prout was clear about his medical goals and their economic significance. He noted that he aimed to provide 'as scientific and accurate a description as possible of the prevalence and distribution of Malaria in Jamaica,' information about the best means to reduce it, as well as current information about the disease for the Jamaican population (W.T. Prout, 'Malaria,' in *Reports of the Twenty-First Expedition of the Liverpool School of Tropical Medicine, Jamaica, 1908–1909*, 476. [afterwards, 'Malaria']). He declared that disease's economic impact demanded 'energetic anti-malarial measures' (Ibid., 493 and 493).

Prout focused his attention on bananas and tourism, the two economic areas of interest to many of Jamaica's political and economic elites and Jones, facts that the British researcher was aware of and that may have shaped his investigations. Prout noted Jamaica's value as a 'health resort' and praised the colony's climate. He noted that the island could become a 'perfect paradise' if Jamaicans would follow his advice about malaria (*DG*, 28 January 1909, 3. See also *DG*, 21 January 1909, 8). Prout also addressed the impact of malaria on the banana industry (Prout, 1910, pp. 376–377; Prout, 'Malaria,' p. 478). He calculated that the disease caused around one-fifth of all deaths in Jamaica and was responsible for approximately one-third of all hospital admissions.[13] Its true cost lay in its impact on the 'general health and physique' of the population and therefore on 'its efficiency for industrial purposes' (Ibid., 493). Malaria made workers inefficient and lethargic, he argued (*DG*, 28 January 1909, 3).

His research convinced Prout that these problems were most severe among the indentured South Asian population and he made a business case for solving the problem (Prout, 'Malaria,' 495–496). Under the indentureship system, almost 40,000 South Asian men and women came to Jamaica as indentured workers between 1841 and the end of indentureship in 1917 (Northrup, 1995, pp. 156–157). Initially, they worked on sugar plantations, but by the turn of the twentieth century, most laboured on banana estates (Look Lai, 1993, p. 185). Political officials and planters argued that the reluctance of former enslaved people to work on the estates meant that these labourers and the indentureship system were vital to maintain estate-based agriculture (Look Lai, 1993, pp. 186–187). Prout pointed to the numbers of indentured workers who were too ill to

[12] Prout also conducted some preliminary research on vomiting sickness, now more commonly recognized as due to consuming underripe ackee.

[13] Prout, 'Malaria,' pp. 492, 495.

work and argued that introducing anti-malarial measures would reduce the incidence of the disease and thus the size of the workforce (Prout, 'Malaria,' p. 496). He was sure that the benefits of this would not 'be disputed by any practical business man' (Prout, 'Malaria,' pp. 496–497). Furthermore, treating these workers with quinine as a prophylactic would make them work more efficiently and prevent them from being treated in hospital, something that would allow the colonial government to save money (*DG*, 28 January 1909, 3).

Although Prout described his methods as influenced by Ross's research protocols, local physicians and planters clearly aided his work and informed his ideas (*DG*, 21 January 1909, 13; *DG*, April 2, 1909). Colonial physicians helped him 'collect' individuals to be examined and provided detailed information about the number of patients in local hospitals suffering from malaria and the number of deaths (*DG*, 2 April 1909). He offered enthusiastic praise for their medical skills and their knowledge of tropical medicine, and he complimented their willingness to follow expert advice (Prout, 'Malaria,' p. 233; 'Island Medical Report, Report for the Year ended 31st March 1910,' p. 211; *DG*, 13 January 1910. 13).

Prout also lauded the 'kindness and hospitality' of Jamaican planters, such as the Scottish-born John Pringle who was a planter and a physician, a dual role that shows the intertwined nature of colonial business and medicine (Prout, 'Malaria,' p. 475). He held key roles in the colony's political and medicine establishment but he was also one of Jamaica's most important banana planters (Sinclair & Fyfe, 1885, pp. 189, 438–439; 1886, p. 547; 1887, p. 579; Cundall, 1920, pp. 95, 201, 222, 240; Ford & Finlay, 1903, pp. 377–379; Musson & Roxburgh, 1897, pp. 415–416; Bryan, 2000, p. 70; Hart, 1954, pp. 214–217; Soluri, 2006, p. 149). Despite this reliance on locals, Prout had advice for them to realize the combined medical and economic goals. He observed that very little had been done in the colony to deal with malaria although its physicians noted the importance of following Ross's approach to the disease (*DG*, 2 April 1909; *DG*, 21 January 1909, 8). He recommended reducing mosquito-breeding areas by filling in swamps and trenches, a recommendation which—despite its high cost—must have gladdened the hearts of the *Gleaner's* editors as this was one of the anti-malarial activities in Sierra Leone they had noted with some envy. In addition, grasses and rubbish should be removed, water barrels and other open containers of water should be screened, kerosene should be placed on top of large pools of water, and individuals should be encouraged to use mosquito nets and to screen their windows (Prout, 'Malaria,' 519–527; *DG*, 28 January 1909, 3). Prout also called for increased use of quinine and recommended that it be made widely and freely available. He called for a public health campaign and public outreach, including the establishment of 'mosquito gangs' and 'quinine dispensers' who would visit sufferers and offer advice and medical referrals (Prout, 'Malaria,' pp. 534–535). In devising these recommendations, Prout drew on his experience in Africa; he recommended that legislation be passed to provide penalties for individuals whose property had mosquito larvae, as 'had been done on the West Coast of Africa' (*DG*, 28 January 1909, 3).

There was a key difference, though, in the policy recommendations for Jamaica and Sierra Leone. Unlike his recommendations for the latter, Prout did not recommend that residential segregation policies be implemented in Jamaica (Prout, 'Malaria,' p. 507). His reluctance may have reflected his unease in calling for actions that contradicted the views of the colonial government. In 1901, Jamaica's superintending medical officer, Charles Mosse, had politely turned down the 'very admirable suggestion' of the Colonial Office that colonial governments consider ensuring that 'all new buildings ... be located away from native quarters.' In explaining this stance, Mosse argued that segregation was 'more applicable' for a place that was 'newly opened up' than it was for a place like Jamaica, 'where it would be difficult to carry out the suggestion' (Chamberlain, Circular to all Colonies, 1 April 1901, no. 3377, CO 854/37, TNA; Charles Mosse to the colonial secretary, 2 July 1901, Jamaica no. 25673, CO 137/620, TNA). Mosse's comments may have reflected his assessment that Jamaica's educated Black and mixed-race population was sufficiently influential so as to prevent the easy introduction of such a policy. An awareness of the resistance to this policy by businessmen like Holt and Africans such as Richard Blaize in Sierra Leone also played a role. A desire to maintain good relations with those in power may also have motivated Prout. In Sierra Leone, he went along with the governor on segregation, and in Jamaica, he supported the local authorities in not advocating for it. In both cases, Prout was eager to cooperate with the politicians.

Prout may have been unwilling to address segregation in Jamaica, but he still drew on an imperial paternalistic discourse that represented non-white peoples as incapable of ensuring their own health without the supervision and guidance of more formally educated Europeans and Euro-Americans. He combined this paternalism with an almost reflexive denigration of non-white colonial inhabitants, echoing long-standing views expressed in the Caribbean and Africa, as we have seen. For example, Prout declared that although some would argue that the 'habits and intelligence of the average coolie' would make using screens in their residences 'difficult if not impractical,' these problems would vanish with 'careful supervision' (Prout, 'Malaria', p. 526). In an apparent effort to use a sanitized version of Jamaican history as an appeal to planter and government action, Prout drew on Edward Long's eighteenth-century *History of Jamaica* to argue that Jamaica's planters had a 130-year history of 'caring' for 'the health of employés' (Prout, 'Malaria', p. 526; Long, 1774, p. 514). Prout's use of the word 'employé' rather than 'slave' and his complimentary words about eighteenth-century planters seem to suggest an unwillingness to offend those who employed him, a pattern seen in his work in Sierra Leone.

Despite these efforts to downplay the compulsory nature of the work that Jamaicans carried out on plantations both in the past and in the early twentieth century, Prout's discussion about the best ways to deal with malaria pointed to the medical benefits associated with a forced labour regime. Part of his rationale for targeting indentured South Asians lay in his certainty that they could be controlled, unlike the rest of the population who could not be compelled to take quinine. On some estates (including one of Pringle's), doses of quinine had been given to Indian men as an 'experiment.' Prout believed that the trial had not been conducted in 'a systematic and thorough manner' and recommended better supervision and record keeping (Prout, 'Malaria', p. 526). A 'quinine register' should be maintained and a 'morning parade' held under the watchful eye of a 'book-keeper,' who should not, he emphasized, be South Asian. Prout maintained that it was 'useless relying on native Indian [South Asian] overseers at first at any rate, until they [began] to appreciate the benefits' (Prout, 'Malaria', pp. 529–530).

Prout's representations of South Asian and Indo-Jamaican labourers as requiring oversight and as appropriate subjects for medical experiments did not seem to have invalidated his recommendations, at least from the perspective of those groups in the colony who had called for anti-malaria initiatives. To them, economic factors seemed more pressing. The *Gleaner*, for example, called for local authorities, especially in key tourist areas, to 'take due heed' of Prout's advice (*DG*, 21 January 1909, 8). After Prout's departure, the *Gleaner* complained about the colonial government's reluctance to follow the researcher's recommendations despite its fulsome praise and its promises (*DG*, 29 January 1909; *DG*, July 24, 1909, 7).

The editors had a point. Several months after Prout left the colony, rather than implementing the researcher's advice, Governor Sydney Oliver called for more studies about the extent of 'fever breeding mosquitoes' in Jamaica and the 'actual conditions that produce[ed] and maintain[ed] them' (*DG*, 29 September 1909, 3). Following Prout's visit and a local outbreak of the disease, the colonial government established its own malaria commission in 1910 and followed up with a number of anti-malaria initiatives, some of which Prout had recommended. These included the beginnings of an education campaign, the passage of a new public health law, training for sanitary inspectors, the sale of cheap quinine in the post office, and government recommendations that district medical officers take on the work of medical officers of health (Prout, 'Malaria in Jamaica', pp. 377–379; Prout, 'Reports of the Twenty-First Expedition of the Liverpool School of Tropical Medicine, Jamaica, 1908–1909,' 471–551; P. C. Cork and E. Jamaica, 'Appendix No. IX. Preliminary Report of the Commissioners on Malarial Fever,' February 1910, in Jamaica, 'Departmental Reports for 1911–1912,' CO 140/241, 1–3). They were short-lived, however. Within a few years, many of these measures seemed to have been abandoned ('First Year's Record of Jamaica's Imperial Association: Report to be Presented Today,' *DG*, 18 February 1919, 13).

Conclusion

The crucial nexus of science, business, and government shaped approaches to malaria control in Freetown and Jamaica. In both places, an approach that would have improved sanitation infrastructure and introduced broader public health initiatives came second to vector control methods. Prout's personal trajectory is also instructive; it shaped his views to conform to business, political, and the LSTM's priorities which ultimately brought rewards. In 1913, asked again about Prout's lectureship, Ross said that he deserved 'a salaried post as Lecturer in Clinical Medicine' as he was one of only a few men in Liverpool who had the necessary experience. As Medical Advisor to the Colonial Office, moreover, Prout 'has done a great deal to help the Colonial Office and its committees and has served the School for many years' (Ross to William Lever, 9 December 1913, LSTM ROSS/113/22/19/3/3). In September of 1913, Prout was duly appointed as a 'Lecturer on Tropical Sanitation' ('Meetings of the Professional Sub-Committee,' 24 September 1913, LSTM

ROSS/113/37/52). Prout's salary was only £100 rather than the £200 that Ross had proposed (RR to Harrison Williams, 7 June 1913, LSTM ROSS/113/37/14/113).

These political and personal intrigues help to explain how scientists coalesced around certain ideas, but despite this frame, local exigencies did shape and colour specific responses as well. Indeed, ideas developed in one context could not be universally applied without modifications, even with the endorsement of powerful groups of capitalists, scientific experts, and colonial officials. The local context shaped policies in a variety of important ways. In Jamaica, labour was key: the rural population on plantations were the focus, whereas in Freetown, the goal was to protect urban inhabitants. Further, segregation was rejected by the colonial government in Jamaica, which helped to shift the recommendations towards other policies. The expedition to Jamaica, moreover, points to the importance of local collaboration. In both places, scientists worked on behalf of elites in government and business; in Jamaica, there was a deep concern for the profits of planters, whereas in Sierra Leone, the elites deemed important to protect were expatriate Europeans. Prout's recommendations conformed both to the politics of the LSTM and Colonial Office but also local political power brokers. In this, he reflected the complicated nature of health policies as they intersected with multiple imperial economic and political ambitions. There were different outcomes in Jamaica and Freetown, but in both places, these interests, coupled with pragmatic adaptation of policies 'on the ground,' played a significant role.

Acknowledgments Three Insight Development Grants from the Social Science Research and Humanities Council of Canada (SSHRC) and several internal grants from McMaster University supported research for this article. With the usual caveats that any errors are our responsibility, we note heartfelt appreciation for colleagues and research assistants who helped gather some of this material, Myles Ali, Mandy Vyce, Ateeka Khan, and Sandria Green-Stewart. And warm thanks for the timely suggestions provided by the attendees at the 2014 conference of the American Association for the History of Medicine, the editors of this collection, and the outsider reviewers.

References

Arnold, D. (1999). 'An Ancient Race Outworn': Malaria and Race in Colonial India, 1860–1930. In W. Ernst & B. Harris (Eds.), *Race, Science and Medicine, 1700–1960*. Routledge.

Austen, E. E. (1899). *Report of the Expedition for the Study of the Causes of Malaria*. Darling & Son.

Bangura, J. (2017). *The Temne of Sierra Leone: African Agency in the Making of a British Colony*. Cambridge University Press.

Birbalsingh, F. M. (1982). The Novels of H. G. DeLisser. *The International Fiction Review, 9*(1), 41–46.

Bryan, Patrick E. (2000). *The Jamaican People 1880–1902: Race, Class and Social Control*. The University of the West Indies Press.

Cole, G. R. (2013). *The Krio of West Africa: Islam, Culture, Creolization, and Colonialism in the Nineteenth Century*. Ohio University Press.

Cundall, F. (1920). *The Handbook of Jamaica for 1920*.

De Barros, J. (2014). *Reproducing the British Caribbean: Sex, Gender, and Population Politics After Slavery*. University of North Carolina Press.

Dumett, R. E. (1968). The Campaign Against Malaria and the Expansion of Scientific Medical and Sanitary Services in British West Africa, 1898–1910. *African Historical Studies, 1/2*, 153–197.

Farley, J. (2003). *Bilharzia: A History of Imperial Tropical Medicine*. Cambridge University Press.

Ford Jos, C., & Finlay, A. A. C. (1903). *The Handbook of Jamaica for 1903*.

Frenkel, S., & Western, J. (1988). Pretext or Prophylaxis? Racial Segregation and Malarial Mosquitos in a British Tropical Colony: Sierra Leone. *Annals of the Association of American Geographers, 78*(2), 211–228.

Goerg, O. (1998). From Hill Station (Freetown) to Downtown Conakry (First Ward): Comparing French and British Approaches to Segregation in Colonial Cities at the Beginning of the Twentieth Century. *Canadian Journal of African Studies, 32*(1), 1–31.

Greenhill, R. G. (1969). The State Under Pressure: The West Indian Mail Contract. *Business History, 11*(2), 120–127.

Harrison, M. (2010). *Medicine in an Age of Commerce and Empire: Britain and Its Tropical Colonies 1660–1830*. Oxford University Press.

Hart, A. (1954). The Banana in Jamaica: Export Trade. *Social and Economic Studies, 3*(2), 212–229.

Holt, T. (1992). *The Problem of Freedom: Race, Labor, and Politics in Jamaica and Britain, 1832–1938*. The Johns Hopkins University Press.

Jalloh, A. (2018). *Muslim Fula Business Elites and Politics in Sierra Leone*. University of Rochester Press.

Jones, M. (2013). *Public Health in Jamaica, 1850–1940.* University of the West Indies Press.

Lambert, D., & Lester, A. (Eds.). (2006). *Colonial Lives Across the British Empire: Imperial Careering in the Long Nineteenth Century.* Cambridge University Press.

Long, E. (1774). *A History of Jamaica or, General Survey of the Ancient and Modern State of that Island.*

Look Lai, W. (1993). *Indentured Labor, Caribbean Sugar: Chinese and Indian Migrants to the British West Indies, 1838–1918.* Johns Hopkins University Press.

Logan Taylor, M. (1902). *Second Progress Report of the Campaign Against Mosquitoes in Sierra Leone.* University Press of Liverpool.

Lumsden, J. (1991). Jamaica International Exhibition, 1891. *Jamaican Historical Society Bulletin, 10*(3–4), 17–21.

Lyons, M. (1992). *The Colonial Disease: A Social History of Sleeping Sickness in Northern Zaire, 1900–1940.* Cambridge University Press.

Macqueen, A. (2004, 2005). *The King of Sunlight: How William Lever Cleaned Up the World.* Corgi Books.

Musson, S. P., & Roxburgh, T. L. (1897). *The Handbook of Jamaica for 1897.*

Neill, D. (2012). *Networks in Tropical Medicine: Imperialism, Colonialism, and the Rise of a Medical Specialty.* Stanford University Press.

Northrup, D. (1995). *Indentured Labor in the Age of Imperialism, 1934–1922.* Cambridge University Press.

Nye, E. R., & Gibson, M. E. (1997). *Ronald Ross: Malariologist and Polymath.* Palgrave Macmillan.

Packard, R. (2007). *The Making of a Tropical Disease: A Short History of Malaria.* Johns Hopkins.

Patton, A. J. (1996). *Physicians, Colonial Racism, and Diaspora in West Africa.* University Press of Florida.

Power, H. J. (1999). *Tropical Medicine in the Twentieth Century: A History of the Liverpool School of Tropical Medicine, 1898–1990.* Kegan Paul International.

Prout, W. T. (1910). Malaria in Jamaica. In R. Ross (Ed.), *The Prevention of Malaria* (pp. 376–381). E. P. Dutton & Company.

Ross, R. (1901). *First Progress Report of the Campaign Against Mosquitoes in Sierra Leone.* University Press of Liverpool.

Ross, R. (1910). *Prevention of Malaria.* E.P. Dutton & Company.

Rush, A. S. (2004). *The Bonds of Empire. West Indians and Britishness 1900–1970.* Ph.D. dissertation. American University.

San Miguel, P. (2011). Economic Activities Other than Sugar. In K. O. Laurence (Ed.), *General History of the Caribbean, Volume IV, The Long Nineteenth Century: Nineteenth-Century Transformations* (p. 124). Macmillan Education/ UNESCO Publishing.

Silva, K. T. (1997). 'Public Health' for Whose Benefit? Multiple Discourses on Malaria in Sri Lanka. *Medical Anthropology, 17*(3), 195–214.

Sinclair, A. C., & Fyfe, L. R. (1885). *The Handbook of Jamaica for 1885–86.*
Sinclair, A. C., & Fyfe, L. R. (1886). *The Handbook of Jamaica for 1886–87.*
Sinclair, A. C., & Fyfe, L. R. (1887). *The Handbook of Jamaica for 1887–88.*
Soluri, J. (2006). Bananas Before Plantations. Smallholders, Shippers and Colonial Policy in Jamaica, 1870–1910. *Iberoamerica, 6*(23), 143–159.
Spitzer, L. (1968). The Mosquito and Segregation in Sierra Leone. *Canadian Journal of African Studies, 2*(1), 49–61.
Taylor, F. F. (1988). The Ordeal of the Infant Hotel Industry in Jamaica. *The Journal of Imperial and Commonwealth History, 16*(2), 201–217.
Taylor, F. F. (1993). *To Hell with Paradise: A History of the Jamaican Tourist Industry.* University of Pittsburgh Press.
Velmet, A. (2020). *Pasteur's Empire: Bacteriology in France, Its Colonies and the World.* Oxford University Press.
Webb, J. L. A., Jr. (2009). *Humanity's Burden: A Global History of Malaria.* Cambridge University Press.
Wilkinson, L., & Hardy, A. (2001). *Prevention and Cure: The London School of Hygiene and Tropical Medicine: A Twentieth Century Quest for Global Public Health.* Kegan Paul.
Will, H. A. (1970). *Constitutional Change in the British West Indies, 1880–1903.* Clarendon Press.
Wise, K. S. (1919). Malaria, The Problem of British Guiana. In *British Guiana Medical Annual for 1919* (pp. 1–28).
Worboys, M. (1988). Manson, Ross and Colonial Medical Policy: Tropical Medicine in London and Liverpool, 1899–1914. In R. Macleod & M. Lewis (Eds.), *Disease, Medicine, and Empire: Perspectives on Western Medicine and the Experience of European Expansion* (pp. 21–37). Routledge.

CHAPTER 5

Leishmaniases in Brazil: A Historical Approach

Jaime Larry Benchimol

Leishmaniasis is the only neglected tropical disease that is still growing, and Brazil is the country in the Americas with the most cases of this disease in its three main forms: cutaneous, mucocutaneous, and visceral (World Health Organization, 2017, 2010). In this chapter, I will show how visceral leishmaniasis became established as a public health problem in Brazil and the state of the art for all the leishmaniases in the 1960s, when they began to appear on the international health agenda.

In Western medicine, production of knowledge on the different illnesses that would be grouped together as "leishmaniasis" in the early twentieth century is associated with European colonial expansion during the eighteenth and nineteenth centuries, as was beautifully framed by Capocci and Cozzoli in the introduction to this book. In India, records of *kala-azar*, a disease that affected the internal organs (particularly the spleen and liver) and came to be known as visceral leishmaniasis, proliferated from the mid-nineteenth century onward, with an upsurge in

J. L. Benchimol (✉)
Casa de Oswaldo Cruz/Fundação Oswaldo Cruz, Rio de Janeiro, Brasil

M. Capocci, D. Cozzoli (eds.), *Empire, Nation-building, and the Age of Tropical Medicine, 1885–1960*, Medicine and Biomedical Sciences in Modern History,
https://doi.org/10.1007/978-3-031-38805-7_5

the extremely deadly epidemics that depopulated various regions and villages in that British colony. The disease was initially associated with malaria or hookworm disease and came to be integrated into the leishmaniasis complex in the early twentieth century, at the same time as dermatological diseases known in various parts of the world by a number of names indicating where they were acquired: Aleppo boil, Biskra button, Bahia button, and so on. "Oriental sore" was proposed in the 1870s as a generic designation to facilitate medical research on the nature of these ailments (Dutta, 2008, 2009; Benchimol & Jogas Jr., 2020).

Leishmaniasis (such a problematic term when used in the singular) challenged the basic Pasteurian concept that each disease has a specific etiological agent, since these diseases with nothing in common were caused by protozoans that were indistinguishable according to the technical resources available at that time and for long afterward (a condition that differed from other protozoan diseases such as sleeping sickness, the object of Amaral's chapter). The leishmaniases became the subject of intense scientific production in various parts of the world. Starting in 1909, Latin American physicians played an important role in this international network, with their research on the singular skin and mucosal manifestations of the disease that had another unique characteristic in the region: it was only acquired in forested areas, not urban centers, as in the Old World. The acceptance of the concept of "American tegumentary leishmaniasis" in the 1930s indicates the projection that Latin Americans achieved in the international network of tropical medicine, including discussions on the role of phlebotomine flies as the probable vectors of this disease complex. But for quite some time, what was unique about the Americas in terms of visceral leishmaniasis was the virtual absence of this form of the disease. One diagnosis was published by a Paraguayan doctor, Luis Enrique Migone Mieres (1913), in an individual who had worked on the construction of the Noroeste do Brasil railroad line. Four years earlier, this same railroad saw an outbreak of "Bauru ulcer," which was then recognized—for the first time in the Americas—as autochthonous cutaneous and mucocutaneous leishmaniasis.[1]

[1] Work by Antonio Carini and Ulysses Paranhos (the director and a research assistant at the Pasteur Institute of São Paulo) and Adolpho Carlos Lindemberg (who headed the dermatology department at Santa Casa de Misericórdia and conducted research at the Instituto Bacteriológico de São Paulo) is analyzed in the study by Benchimol and Jogas Jr. (2020, pp. 46–47). The case of visceral leishmaniasis observed in Paraguay and two others subsequently diagnosed in Argentina were seen as isolated events in stark contrast with hundreds of cases of tegumentary leishmaniasis described by Latin American doctors, suggesting that the dermatological disease could also attack the viscera.

Visceral leishmaniasis only erupted as a public health problem in 1934. The Yellow Fever Service had just created a laboratory to analyze liver samples collected from people who died of suspicious fevers by hundreds of viscerotomy posts scattered throughout Brazil. The pathologist Henrique Penna identified *Leishmania* in slides that tested negative for yellow fever, and, as a result, 41 deaths were suddenly related to visceral leishmaniasis (Penna, 1934).

With these *post mortem* case reports, Evandro Chagas (1905–1940), the son of Carlos Chagas, found the case diagnosed in a living patient in the northeast of Brazil (Chagas, 1936a). In a note published soon afterward (March 1936b), this researcher at the Oswaldo Cruz Institute presented visceral leishmaniasis in Brazil as possibly differing from the forms observed in the Old World, even though it was impossible to differentiate the *Leishmania* found in Brazil, *Leishmania donovani*, the agent of *kala-azar* in India and other parts of Asia, and *Leishmania infantum*, the parasite described by Charles Nicolle in 1908 as the pathogen of infantile visceral leishmaniasis in the Mediterranean region (Chagas, Oct., 1936b. See also Chagas et al., 1937. On Evandro Chagas's work on visceral leishmaniasis, see Deane, 1986; Benchimol et al., 2019).

As head of an official commission to study visceral leishmaniasis,[2] Evandro Chagas visited other places in the northeast, but soon moved his research to Pará, in northern Brazil.[3] His investigation continued to be guided by reports from Yellow Fever Service pathologists. Evandro Chagas made great efforts to demonstrate *a priori* the alleged autochthony of "American visceral leishmaniasis." He sought to repeat the feat of those who had succeeded in establishing the concept of American cutaneous leishmaniasis as well as his father Carlos Chagas, who discovered the famous American trypanosomiasis in 1909. Evandro Chagas was certainly influenced by the scientific effervescence caused by the discovery of syl-

[2] The Commission included three researchers from the Oswaldo Cruz Institute: Aristides Marques da Cunha, Gustavo de Oliveira Castro, Leoberto de Castro Ferreira, and the Argentine Cecílio Romaña. Evandro Chagas had financial support from the Oswaldo Cruz Institute, the Rockefeller Foundation, and a Brazilian businessman, Guilherme Guinle. Evandro Chagas's relationship with these sponsors of his activities is analyzed in Barreto (2012).

[3] In Belém, Pará, other professionals were soon incorporated into his team: the English nurse Agnes Stewart Waddel (who would become his second wife), Leônidas Gladstone Deane (from Pará), Felipe Nery-Guimarães, and Maria von Paumgartten (who would later marry Leônidas Deane).

vatic yellow fever, with the participation of the pathologist Henrique Penna (Soper et al., 1933). A similar effervescence is described by Silva's chapter on sylvatic plague.

Chagas and his collaborators stated that the disease occurred in individuals of different ages (especially those cases diagnosed posthumously through viscerotomies), while Mediterranean visceral leishmaniasis was mainly seen in children. In the Americas, the disease was only found in rural zones bordering wild areas, unlike *kala-azar*, which manifested as intense urban epidemics in India. Dogs were the main reservoirs in the Mediterranean zone, but could not play this role in a disease as sparse as the American one. Only wild animals acting as primary reservoirs could explain the epidemiological profile of American visceral leishmaniasis.

In incriminating a new species, *Leishmania chagasi*, as its causal agent, Evandro Chagas followed the example of Gaspar de Oliveira Vianna (1911), a pathologist at the Oswaldo Cruz Institute who had described *Leishmania braziliensis*, one of the many explanations offered for the unique characteristics of tegumentary leishmaniasis in the Americas.[4] Vianna's and Chagas's theories were bolstered by the inclusion of nationalist ideologies in the medical debate, since an autochthonous protozoan species strengthened the symbolic capital of Latin American researchers within the international scientific network of tropical medicine.

The validation of *L. braziliensis* is also related to the interpretation of Pre-Columbian ceramics that show human figures with lesions on the nose and mouth, considered proof that mucocutaneous leishmaniasis originated in the region. But the interpretation of these archaeological findings was not consensual; from the late-nineteenth century, major controversy emerged among archaeologists, ethnologists, naturalists, and physicians from the Old and New Worlds, all disagreeing about the pathologies or practices suggested by these ceramics as well as mummies, bone remains, and other Pre-Columbian artifacts. To support their claims, these specialists used photographs and other iconographic representations from their own respective time periods to illustrate the effects of each disease being discussed. This discussion has significant heuristic value, since it reveals the complex of pathologies from which tegumentary leishmaniasis had to be untangled as it was being posited as a new disease native to the Americas, as shown by Benchimol & Jogas Jr., (2020).

[4] The original name (*Leishmania brasilienses*) was altered to *L. braziliensis* by Vianna himself in a 1914 article, but the name *Leishmania brasiliensis* (with the *s*) is still found in many scientific articles up through the 1960s.

It is also important to remember the prominence Gaspar Vianna achieved after his 1912 discovery that emetic tartar, a trivalent antimony compound, was an effective treatment for leishmaniasis despite its toxicity.[5] The drug was soon adopted in many countries and colonies and had a strong impact on the mortality rates of *kala-azar* in India.

Evandro Chagas's theory about American visceral leishmaniasis largely stemmed from the fact that his research focused on the Amazon, where this form of the disease was scarce until the 1980s compared to northeastern Brazil, or even compared to cutaneous and mucocutaneous leishmaniasis, which had been intense in the north of the country since the beginning of the rubber economy. Evandro Chagas chose the Amazon because it was the only region where his political negotiations were successful, resulting in the creation of the Instituto de Patologia do Norte (Northern Pathology Institute, IPEN) in Belém, Pará to house his studies.[6]

The first living cases of visceral leishmaniasis in the Amazon were found in a village called Piratuba, and a few more were later identified in three other municipalities in Pará: Marapanim and Soure, on the coast, by viscerotomy, and Mojú, by clinical assessment.[7] These cases corroborated the idea that the disease did not occur in towns or floodplains (namely areas with major waterways or flood zones), but rather on *terra firme*, forested areas with dry land and large concentrations of wild mammals. In Pará, natural infections were found in seven dogs and one cat. Granulations in a wild rodent (a tree rat, *Phyllomys* ssp.) suggested *Leishmania*, but this finding could not be confirmed. The most common fly species suspected of transmitting visceral leishmaniasis in the region were *Phlebotomus longipalpis* (*Lutzomyia longipalpis*) and *Phlebotomus intermedius* (*Lutzomyia intermedia*), but their role was not conclusively proven despite knowledge accumulated since the 1920s on phlebotomine sandflies as vectors of

[5] "Tratamento da leishmaniose pelo tártaro emético" [Treatment by emetic tartar] was a communication presented in a session of the Brazilian Society of Dermatology during the seventh Brazilian Congress on Medicine and Surgery, in Belo Horizonte, the capital of Minas Gerais, in April 1912 (Vianna, 1912). In India, after 1922, emetic tartar was replaced by *urea stibamine*, a pentavalent antimony developed by Upendranath Brahmachari. Other antimonial compounds such as Fuadin were produced by pharmaceutical companies like Bayer. On this topic, see Brahmachari (1940), Marsden (1985), and Barbeitas (2022).

[6] On this, see Gualandi (2013), Barreto (2012), and Deane (1986).

[7] A total of 14 cases were diagnosed by viscerotomy in Pará between 1932 and 1940; Evandro Chagas and his team identified eight living cases.

cutaneous and mucocutaneos leishmaniasis in the New and Old Worlds, as shown by Dedet (2005).[8]

Evandro Chagas and his collaborators established close connections with Argentinian physicians. In 1935, Chagas participated in the ninth Meeting of the Argentine Society of Northern Pathology (SAPRN) in the city of Mendoza, which honored his father Carlos at a crucial moment in the history of his discovery. Chagas disease had been thrown into limbo as an illness with a supposedly limited magnitude, but its true epidemiological extent was revealed in a type of conjunctivitis known as Romaña's sign, discovered by a member of this society, Cecilio Félix Romaña. He joined the commission headed by Evandro Chagas and was part of the discovery of American visceral leishmaniasis.

Chagas's most prestigious international ally was Saul Adler, a Belarus-born physician educated in England who started his work at the Hebrew University in Jerusalem in 1923.[9] With the entomologist Oskar Theodor, he studied recognized or disputed species of *Leishmania*, developed ingenious serological methods to differentiate them, and investigated the known and alleged vectors of the parasite and zoogeography of leishmaniasis in the Mediterranean region. There was already strong evidence indicating the role of phlebotomine flies in transmitting the disease. In 1921, at the Pasteur Institute in Algeria, Edmond and Étienne Sergent and their colleagues demonstrated that *Phlebotomus papatasi* was able to transmit "Biskra button," the name for cutaneous leishmaniasis in that Algerian city: they produced a skin lesion in a volunteer by inoculating crushed naturally infected flies.[10] Using a similar technique, Henrique Aragão of

[8] The blood of an infected dog was consumed by females of *P. longipalpis*, and *Leishmania* identical to those in the protozoan cultures were found in two of these insects (Ferreira et al., 1938). *Phlebotomus intermedius* was also infected using animals with leishmaniasis. And it was with this fly that Agnes Chagas and Wladimir Lobato Paraense obtained their first positive results for transmission of visceral leishmaniasis to a hamster via *Phlebotomus* (Paraense & Chagas, 1940; Chagas, 1939). The production and circulation of knowledge about the possible vectors of leishmaniosis is the subject of a chapter by Benchimol and Jogas Jr. (2020, pp. 145–189). An important reference on phlebotomine flies in Brazil is the book edited by Rangel and Shaw (2018).

[9] Adler assumed the directorship of the Department of Parasitology, which was later incorporated into the Hadassah Medical School, opened soon after the State of Israel was created in 1948. Here my main sources are Shortt (1967) and Telkes (1998).

[10] Sergent et al. (1921, 1926). Authors like Dedet (2007) and Théodoridès (1997) see this as the initial evidence that phlebotomines were the vectors of leishmaniasis. Killick-Kendrick (2013) emphasizes the work done by Swaminath, Shortt, and Anderson (1942).

the Oswaldo Cruz Institute implicated *Phlebotomus intermedius* (*Lutzomyia* [*Nyssomyia*] *intermedia*) in the transmission of *L. braziliensis*.[11] And Adler and Theodor likewise transmitted *L. infantum* to humans via *P. papatasi*,[12] but neither the Europeans who studied leishmaniasis in the Mediterranean region nor the British and Indians investigating *kala-azar* in Asia were able to experimentally transmit the parasite to human volunteers through bites from flies that had previously fed on patients with the disease (the decisive proof).

It was only in 1942 in Assam (one of the Indian states most affected by visceral leishmaniasis) that Henry Edward Shortt and his team proved that humans contracted *kala-azar* after being bitten by *Phlebotomus argentipes* previously infected on sick patients. As Killick-Kendrick (2013, pp. 134–135) and Desowitz (1991, pp. 17–18) explain, this was largely due to a fundamental technical innovation. Chagas, Adler, and other researchers followed methods adopted to study the malaria–mosquito relationship by successive blood meals, but in *Leishmania* this produced noninfectious flies. However, a diet of raisins (i.e., sugar) following an infectious blood meal surprisingly caused the flagellates of *Leishmania* to multiply sharply in the body of the insect.

In a study published shortly before the conclusion of this research, Adler (1940a) made an interesting summary of the frustrated attempts to experimentally infect humans; the most recent at that time were those of Evandro Chagas and his colleagues.[13]

In September 1939, Adler sent Evandro Chagas an article that was published in the journal of the Oswaldo Cruz Institute (Adler, 1940b). "So far it has been impossible to distinguish *L. chagasi* from *L. infantum* by any laboratory test,"[14] wrote Adler (1940b, p. 173), but he proposed other arguments to justify this distinction. First, in the Old World, distribution of *L. infantum* and *L. donovani* was "rigorously confined" to

[11] Aragão published a note on this in 1922 (pp. 129–130) and a more comprehensive work in 1927 (pp. 177–186). Another species described by Lutz and Neiva (1912), *Phlebotomus longipalpis* (today known as *Lutzomyia longipalpis*), was later associated with the transmission of visceral leishmaniasis by Evandro Chagas and his team.

[12] The experiments consisted of crushing naturally infected flies and introducing this material into skin lesions created on human volunteers. Adler, Theodor (1925a, b).

[13] Adler cited Cunha, Chagas, "Estudos sobre o parasito" [Studies on the parasite], part 3 (pp. 329–337) of the report by Chagas et al. (1937) on work done during 1936.

[14] In this and other citations of texts from non-English languages, a free translation has been provided.

Phlebotomus species of the *major* group that were not found in the Americas. The synonymity between *L. chagasi* and *L. infantum* assumed that the latter had been exported to the Americas. However, Adler argued that the existence of *L. braziliensis* in pre-Columbian America (based on the archaeological findings mentioned above) and of a related species of the parasite in the Old World (*L. tropica*) when there was no communication between the two continents indicated the potential antiquity of *L. chagasi*. Export of *L. infantum* to the Americas was considered unlikely because, in his own words, "Up to the present there have been no proved instances of the formation of new foci of *L. infantum* even in the Old World itself 'where the disease is remarkably static in its distribution'" (Adler, 1940b, p. 174).

On November 8, 1940, Evandro Chagas died at age 35 in a plane crash in Guanabara Bay. Studies on visceral leishmaniasis lost momentum in Brazil, and at the same time, the first major epidemiological survey on American tegumentary leishmaniasis was begun by Samuel Barnsley Pessôa (1898–1976), a parasitologist at the recently inaugurated University of São Paulo.

In the inland regions of São Paulo state where the disease was endemic, the activities carried out by the Leishmaniasis Study Commission created through an agreement between the Parasitology Department of the São Paulo School of Medicine and the São Paulo State Department of Health were prolonged for two and a half years (1939–1941) until they were abruptly ended by the state government, largely because of the problems these activities caused for large rural landowners, especially those who divided up part of their properties to create smaller farms or urban settlements in areas that had already been cleared. The epidemiological investigation led by Samuel Pessôa showed that the incidence of the mucosal forms of the disease was much higher than imagined. It also showed that the share of women among leishmaniasis victims had been hidden by the fact that it was mostly men seeking to support their families who went to the larger cities to seek medical care, and only there the disease was registered. The Commission brought care to the rural areas and treated more than nine thousand patients with cutaneous lesions or severe mucosal lesions. Promising experiments were conducted with a vaccine made from dead cells of various strains of *L. braziliensis* (see Benchimol & Jogas Jr., 2020, pp. 193–239).

Of the 34 species of phlebotomine flies identified in the state of São Paulo, the Commission confirmed that only three (*Phlebotomus whitmani*,

P. pessoai, and *P. migonei* [*Lutzomyia whitmani* Antunes & Coutinho, 1939; *Lutzomyia pessoai* Coutinho & Barretto, 1940; and *Lutzomyia migonei* França, 1920]) seemed to play an important role in transmitting cutaneous and mucocutaneous leishmaniasis. These species accounted for nearly all the phlebotomine fauna in the zones with high and low endemicity, which were more precisely demarcated.

Besides nearly 60 articles appearing in various domestic periodicals, Samuel Pessôa and his colleague Mauro Pereira Barretto published *Leishmaniose tegumentar americana* in 1948; this award-winning book became a classic work on this topic and decisively helped solidify this long-lasting concept, particularly among Latin American specialists.

During the postwar period, developmentalist nationalism was embraced by many teachers and researchers in the traditional medical schools and those which were being created around Brazil, as well as by sanitarians who found promising professional prospects in the health services created after the 1930 revolution that brought Getúlio Vargas to power. Mário Pinotti, the head of the National Malaria Service, led the team responsible for the health program of Juscelino Kubitschek, elected president in 1956 on the promise of "50 years of progress in 5 years of government" (see Hochman, 2009). The living conditions of rural workers were expected to improve with eradication or control of endemic diseases in the interior of the country using specific strategies tailored to the biological and social peculiarities of each disease and the availability of antibiotics, insecticides, vaccines, and other prevention or control techniques. Also influential were the priorities established as international health agencies played an increasingly important role in decisions by the governments of Brazil and other countries.

While diagnoses of tegumentary leishmaniasis increased in number, there were few *in vivo* cases of visceral leishmaniasis. Since the first diagnosis in 1913, 34 cases had been recognized in living patients in Brazil and only 35 in the Americas until 1953. Viscerotomies indicated 314 more widespread deaths between 1934 and 1950 (Deane & Deane, 1955a).

But in 1953, an epidemic broke out in northeastern Brazil that dramatically changed this situation: in just five years (1953–1957), the total number of *in vivo* cases jumped from 34 to 1832, the majority (81.38%) in the state of Ceará. In the Americas, the number rose to 2179, but more than 98% of the total were found in northeastern Brazil (Deane, 1958).

At that time, a devastating drought punished the region, and thousands of people left in search of food, shelter, and work. In the press and among the doctors and the people, there was much talk about "*calazar.*" Contemporary investigations showed that it had gone unnoticed for quite some time. The few previous diagnoses did not reflect the rarity of the disease, but rather the lack of medical assistance available to the rural population and scarcity of knowledge among the doctors working in the interior. A campaign against visceral leishmaniasis began in Ceará under the leadership of Joaquim Eduardo Alencar, one of the founders of the Ceará School of Medicine in 1947, where he taught parasitology. After visiting the stricken region, Samuel Pessôa sent two of his assistants there, Leônidas Deane (1914–1993) and his wife Maria von Paumgartten Deane (1916–1995), both former members of Evandro Chagas's team.

In Ceará, they found a fox (*Lycalopex vetulus*) infected with *Leishmania* (Deane & Deane, 1954). They thought they had been the first to find a wild animal with visceral leishmaniasis, but soon learned that three Russians had already verified in Central Asia (Tajikistan) that the jackal (*Canis aureus*) also played this role.[15] While this confirmed the wild reservoir that Evandro Chagas had sought, the rest of the observations made by Leônidas and Maria Deane and local researchers seriously undermined the theory proposed in the 1930s. Now the disease did not depend on the forests; it could occur in urban areas, and even in the rural zones where it predominated it had a "focal" character. Urban transmission was confirmed by individuals and dogs infected in the cities.

Most of the rural cases were concentrated in areas known as *boqueirões* and *pés da serra*, narrow valleys between mountains or in their foothills. In the arid *sertão* and verdant highlands, infection was infrequent and cases were sparse and sporadic. The endemic area was soon extended after investigations by local physicians in the northeast and north of the country; initially, the Jaguaribe River valley, where a recent campaign against *Anopheles gambiae* had taken place (Lopes, 2019), and later the banks of the Piancó River in Paraíba (Alencar et al., 1956; Alencar, 1962).

And how did the campaigns combat *kala-azar*? As in other parts of the world, through DDT spraying.[16] Phlebotomine flies disappeared from the

[15] Latyshev, Kryukova, and Povalishina (1951). This and other articles by Soviet researchers were reviewed and/or commented on by the British parasitologist Cecil Arthur Hoare (1954).

[16] Research was important to calibrate the use of DDT against *Phlebotomus longipalpis*: Deane, Deane, Alencar (1955). At least until 1960, the use of this residual insecticide in Ceará continued to be experimental in nature, since it was applied in one group of 14 municipalities but not in another in order to evaluate the relative efficacity of other prophylactic measures (Alencar, 1961).

treated houses, but the active duration of the insecticide was shorter in animal shelters outdoors and it proved useless in peri-domicile areas. Intradomiciliary transmission could be reduced because *P. longipalpis* attacked mostly at night, when the human population was inside their homes, but outdoor transmission was the Achilles' heel of this prophylactic strategy. The antilarval measures that were so important in yellow fever and malaria were not applicable against this disease since larvae of *P. longipalpis* had been found in the soil under trees, not in aquatic environments. In fact, the breeding habits of this and other species of sandflies were still poorly understood.

Spraying was done by the National Malaria Service according to the habits of *Anopheles*, which were not the same as *Phlebotomus*. "Spraying was not done to the necessary extent," complained Alencar (1961, p. 178), "nor was it done at the suggested time, preceding the rains, after which the density of phlebotomine flies increases, and with it contagion intensifies."

The campaign against visceral leishmaniasis in the northeast of Brazil also involved the discovery and elimination of canine cases. Apparently, only Brazil and China may have killed dogs on such a large scale in order to control leishmaniasis. In fact, in his habilitation thesis on canine leishmaniasis, Alencar (1959) used the Chinese experience as his main reference, for there the strategy against *kala-azar* relied mainly on destroying the intermediate host of *L. donovani*. His main sources were from the study of Chung (1940), who advised elimination of sick dogs, and an article on New China's achievements in treating and preventing *kala-azar*, which was presented by Lu et al. (1955) at a scientific meeting sponsored by the USSR Academy of Medical Science and the Ministry of Public Health of the Uzbek Soviet Socialist Republic in September 1954 at Tashkent.

Alencar and the physicians who worked on the Brazilian campaign against *kala-azar* had no doubts that the socioeconomic conditions of rural workers in northeastern Brazil were largely responsible for the prevailing epidemics. A statement attributed to Alencar sums up the situation perfectly: "Leishmaniasis is a disease of dogs and of those who lead a dog's life." Consequently, prophylaxis needed to include improving the living conditions of affected populations. In practice, however, actions related to the human populations were limited to systematic treatment with pentavalent antimonial drugs, chiefly Glucantime (Rhône-Poulenc) and Solustibosan (Bayer), which were provided to hospitals and health centers at no charge.

In 1958, the Ministry of Health under Juscelino Kubitschek promised to eliminate five endemic diseases: yaws, Chagas disease, goiter, trachoma, and leishmaniasis. That same year, the fifteenth Pan American Sanitary Conference in Puerto Rico approved a resolution declaring Brazil, the Panama Canal Zone, and nine other countries free of *Aedes aegypti*, the urban vector of yellow fever. Also in 1958, Brazil officially joined the worldwide campaign to eradicate malaria. These events marked an era of sanitary optimism, which a decade later began to give way to the uncertainties of the present day. Meanwhile, activities against rural endemic diseases rolled out throughout Brazil. Studies at that time showed that the emigration of rural workers from the northeast increased the incidence of the leishmaniases in other regions of the country. The number of cases diagnosed in the interior grew alongside those in the big cities. Awareness also spread that isolated cases of visceral leishmaniasis required investigation, since they were likely to be linked to new endemic areas.

These epidemiological surveys were made possible by transformation in the public health apparatus at the federal and state levels and by strengthening the institutional network to house studies on the leishmaniases and other rural endemics. The medical schools in São Paulo and Ribeirão Preto (SP) (the latter created in 1947) and the Oswaldo Cruz Institute in Rio de Janeiro were the main nodes of this network, which included the Institute of Preventive Medicine at the University of Ceará, the Evandro Chagas Institute, the Special Public Health Service (SESP), and three research centers established in the late 1950s by DNERu (the National Department of Rural Endemic Diseases): the Bahia Research Center, in Salvador (later called the Gonçalo Moniz Research Center), the Belo Horizonte Research Center (which would be called the René Rachou Research Center), and the Aggeu Magalhães Research Center in Recife. Work in these institutions (which since the 1970s were subordinated to Oswaldo Cruz Foundation) focused mainly on schistosomiasis, Chagas disease, and malaria, but visceral leishmaniasis gained importance in studies involving cooperation from the universities of Pernambuco, Bahia, and Minas Gerais.

A landmark in the consolidation of this research network was the *Kala-azar* Seminar held in Salvador, the capital of Bahia, in 1960. Experts interested in this endemic disease (but who also worked with others) met for the first time, and on the first day of the event, the attendees voted to create the Brazilian Society of Tropical Medicine (see Benchimol & Jogas Jr., 2020).

Tropical Medicine Within the Context of the Cold War

Tropical medicine also gained renewed importance internationally as the traditional diseases in this category began to pose significant obstacles (and resulting byproducts) to development policies that targeted countries like Brazil to lift them out of their "underdeveloped" status; more notably, it was important to stop these countries from escaping their orbits around the capitalist world through popular rebellion propitiated by poverty and disease.

Let us put this into context with a quick sketch of the background at this time.

In the years following the end of World War II during the lead-up to the Cold War, the institutional arrangements and political alignments, which characterized the field of health, began to change and diversify. The establishment of a bipolar geopolitical order exacerbated antagonisms and limited how multilateral or global health concepts could be applied. The bilateral approach adopted by the United States during World War II through the Good Neighbor Policy was further solidified with the creation of or encouragement to create different health institutions and services. Notable in Brazil were the maintenance of the Special Public Health Service as a strategic agency for American and Brazilian interests in the north and northeast regions of the country (Campos, 2006) and initiatives by the Rockefeller and Kellogg Foundation and the Pan American Sanitary Organization (later called the Pan American Health Organization, PAHO) to organize or reformulate medical education in Brazil and other Latin American countries based on the American or Flexnerian model (Edler & Pires-Alves, 2018; Kemp & Edler, 2004; Cueto, 2015). This model valued the connection between research, teaching, and clinical practice and attributed specialized technologies and practices the power to "solve" health problems in the region. This belief reached its zenith in the campaigns to eradicate or control malaria and other diseases through insecticides or chemical therapy in verticalized activities. The health optimism that reigned during the postwar period was important in mobilizing the medical field and one of the keys to the multilateral linkages that led to the design of international health in this period when two agencies emerged. The Pan American Sanitary Organization (later called PAHO) and the World Health Organization (WHO) played an important role in reflections on health and development and in proposing agendas to confront diseases and other health problems at global and regional levels. Starting in 1947 under the leadership of Fred L. Soper, who was highly regarded for his work on campaigns against yellow fever and malaria in the region, the PAHO placed great emphasis on training human resources for health and proposing strategies for the governments of the Latin American countries to face common health problems. This commitment can be seen in the organization of the Pan American Health Conferences, which from 1947 were dedicated to the theme of international health and its challenges (Lima, 2002; Cueto, 2015).

The World Health Organization fostered discussions on health and development, especially during the management of the Brazilian Marcolino Gomes Candau, who in 1953 was elected the organization's second director general and was reelected to serve four consecutive mandates, until 1973. The withdrawal of the Soviet Union and its allies from the UN system (including WHO) between 1949 and 1956 reinforced the perception that this system was dominated by the interests of the United States (Brown et al., 2006). In addition to articulating major campaigns for disease control or eradication and planning regional and multilateral health strategies, WHO played a fundamental role in reformulating the system of validating medical and health knowledge through its world health assemblies, international conferences, and reports and opinion papers produced by experts it recruited to present their opinions on specific topics.

At this point in the postwar period, ideas about tropical diseases and medicine's ability to transform underdeveloped countries like Brazil were strongly influenced by an economist vision of health planning, based on analysis by authors such as Charles-Edward Amory Winslow (1955) and Gunnar Myrdal (1952) about the vicious circle of poverty and disease. According to these authors, socioeconomic development could only be achieved by combating the large-scale diseases that impeded it.

On March 13, 1961, US president John F. Kennedy made a famous speech at the White House before Latin American ambassadors introducing the Alliance for Progress, a development assistance plan for Latin America that would be formalized five months later when the United States and 22 other nations in the Americas (including Brazil) signed the Charter of Punta del Este. The notion of inter-American cooperation had been proposed in 1958 by Juscelino Kubitschek: the so-called Pan American Operation (OPA) was based on the idea that only by eliminating poverty on the continent could communism be contained and democracy expanded. The victorious socialist revolution in Cuba in January of 1959 convinced the formulators of hemisphere policy in Washington that they needed to be more sensitive to increasing demands for economic development and social progress from Latin America. The traditional policy based on the assumption that private capital alone would permit economic growth in the region had faltered (momentaneously, as we can see today).

The Charter of Punta del Este recognized the urgent need for social, economic, and political transformations in the Americas. The short-term goals of the Alliance for Progress were to bring about changes in the daily life of the Latin American masses who faced challenges related to housing, clothing, nutrition, and literacy and in the longer term involved reforms that would ensure socioeconomic development and boost quality of life in each country through national development plans, economic integration, stabilizing raw material prices, and financial restructuring. In the meeting held in Punta del Este, Uruguay, in August 1961, the United States committed to providing a substantial portion of the 20 billion dollars that Latin America would need to complement with its own resources over the next ten years.

Throughout this decade, which was marked by a succession of military coups that shook the continent, the Alliance for Progress was heavily criticized by experts who attacked its weak structure and unrealistic goals as well as liberal and leftist sectors that viewed it as an instrument of American economic and strategic interests.

In 1962, the year of the Cuban missile crisis and exacerbation of the Cold War, a comprehensive study on tropical diseases and the resources for their study and control was completed in the United States.[17] The study, which was sponsored by the National Academy of Sciences and the National Research Council, had support from the most important American agencies, including the Rockefeller Foundation. Its Consultant Committee, chaired by Albert B. Sabin, was a team of heavyweights that brought together health, industry, and business. The arguments justifying the relevance of the study clearly express their concern with preserving a power with imperial ambitions from the dangers that threatened it: decolonization movements, revolutions, and communism.

Nearly half the world's population was in the tropics, with hundreds of millions of people living in poverty. The tropical world was seen as a

[17]There are two versions, a complete 540-page version (*Tropical Health: A Report on a Study of Needs and Resources*) and a summary with 121 pages. They will be referred to here as the National Academy of Sciences—NAS (1962a, b).

powder keg, ready to explode. The relationship between the old metropolises and their colonies, through services and institutes with interests and jurisdiction limited to the areas of each metropolis, had lost its reason for being. The New World order, established under the leadership of the United States, demanded new roles for this country on a global scale.

The authors of *Tropical Health: A Report on the Study of Needs and Resources* stirred up controversy with a Malthusian train of thought that did not have a favorable view of "premature" improvements in public health in the tropical regions, since they would intensify already rapid population growth far beyond what the local economies could bear. For Sabin and his peers, control of widespread diseases would lead to an increase in the proportion and productivity of workers within the population as a whole, giving them "more strength and ambition" to explore new lands or recover those that had been abandoned or considered impenetrable because of disease (NAS, 1962a, p. IX). I note here that this is precisely the factor that was most conducive to the spread of leishmaniasis and diseases that would later be classified as emerging.

Leishmaniases: State of Art

To analyze the state of the art in relation to leishmaniasis, Leônidas Deane of the medical school at the University of São Paulo was consulted, along with Marshall Hertig of the Gorgas Memorial Laboratory (Panama) and Philip Edmund Clinton Manson-Bahr, a British physician then connected to the Department of Health in Nairobi, Kenya.[18] But before we examine their opinions, we need to add another initiative to this scene: in 1960, the World Health Organization commissioned Percy Cyril Claude Garnham, a professor at the London School of Hygiene and Tropical Medicine, to perform a study on the problems related to leishmaniasis, amebiasis, the trypanosomiases, and toxoplasmosis on a global scale, diseases which had previously been underestimated by that agency. One notable aspect of his report regarding the leishmaniases is the contrast between the fixed

[18] The list of consultants also included P.C. Sen Gupta of the Calcutta School of Tropical Medicine in India, but the book does not contain his opinion. A summary of the opinions produced by the rapporteur Willard H. Wright can be found in NAS (1962b, pp. 59–60) and in NAS (1962a, pp. 376-377). The opinion of Leônidas M. Deane is found in 1962a (p. 505), of Marshall Hertig in 1962a (pp. 505–508), and Philip Edmund Clinton Manson-Bahr (1962a, p. 508).

epidemiological scenario described by Adler in the 1930s and its fluidity in the 1960s.

Observations in Russia and Iran had demonstrated the existence of two clinical forms of cutaneous leishmaniasis, "dry" and "wet," which were epidemiologically and serologically different. The classic form was dry, in the cities; the "wet" type was more severe, and occurred in rural areas, steppes, and deserts.[19] From the Mediterranean basin, especially Italy and North Africa, cutaneous leishmaniasis expanded its territory. Each year thousands of cases occurred in Iraq, Iran, Lebanon, Syria, Russia, Pakistan, and India. According to Garnham, slow expansion through the Sahara spread cutaneous leishmaniasis to Sudan, Ethiopia, French West Africa, Nigeria, and Ghana. The equatorial forest seemed to block the advance to the south, but it was critical to determine whether there was danger that the disease would spread widely across Africa (Garnham, 21 June, 1960, pp. 12–13).

In visceral leishmaniasis, Garnham distinguished the classic form (*kala-azar*), the Mediterranean or "infantile" form, forms from Sudan and East Africa, and also "aberrant" leishmaniasis. The epidemiological characteristics differed, and the vectors and reservoirs of the causal agent were still largely unknown. *Kala-azar* mainly occurred on the Indian subcontinent—India and Pakistan, the latter country recently dismembered from the former after the end of British colonialism in the region. The disease prevailed in other parts of Asia (Bhutan, China, Turkistan, Russia), in the Middle East, and on the Mediterranean coast and islands. It was also spreading through Africa; it had reached Kenya via Egypt and Sudan, northern Nigeria and French West Africa through the Sahara. While the total number of cases in the Americas soared impressively from 35 to more than 2179 cases in the 1950s, this surge pales in comparison with the

[19] Garnham (21 Jun. 1960, p. 13). The classifications *minor* and *major* were proposed in 1915 by a mission organized to study the tropical diseases of men and animals in Turkmenistan, which at that time was a province of tsarist Russia. The expedition found that the lesions called by different local names and seen by doctors in the region as different diseases represented a single illness (Yakimoff & Schockov, 1915). In the 1940s, three Soviet researchers (Latyshev, Kryukova, and Mirzoian) justified the existence of two "races" of *Leishmania tropica* from clinical and epidemiological differences in illnesses they produced and immunobiological differences. Kryukova showed that the two parasites could be differentiated by inoculation into gerbils and white mice: these animals were resistant to *L. tropica minor* (the dry type), but the agent of the wet type (*L. tropica major*) produced infections. References to this research can be found in the invaluable work by Heyneman, Hoostraal, and Djigounian (1980).

fragmentary data compiled by Garnham: in hospitals and dispensaries in West Bengal, there were 38,848 cases in 1951 and 10,000 cases in Sudan in 1959 (Garnham, 1960, p. 11). In Kenya, a smaller epidemic in 1960 caused about 1000 deaths. Visceral leishmaniasis became a major problem in Sudan and in East Africa at a surprising speed, and there were fears that the disease would spread throughout sub-Saharan Africa.

The experts consulted for the American study agreed that visceral leishmaniasis had diminished after DDT sprayings as part of the malaria eradication programs, but they feared that sandflies and the diseases they transmitted would reappear when spraying ceased. It is important to recall that the international campaign to eradicate malaria was already foundering, in large part due to the DDT resistance seen among *Anopheles* mosquitoes. Many initiatives were taken at the time to evaluate the true dimension of this problem and to find alternatives. Richard Fay, of WHO's Environmental Sanitation Division, asked Leônidas Deane to investigate the susceptibility of *Phlebotomus longipalpis* to DDT and Dieldrin. Experiments done by a technician at the Belo Horizonte Research Center under Deane's supervision led to the conclusion that these insecticides remained effective in destroying the main vector of visceral leishmaniasis in the Americas, a fact that "naturally removes from the problem … the urgency with which it presents itself in the case of anophelines" (Falcão, 1963, p. 411).

Hertig proposed systematic control in the Old and New Worlds, but Manson-Bahr thought it unlikely that DDT would affect the incidence of visceral leishmaniasis in East Africa; in fact, it increased in Sudan after some malaria control activities. The British researcher believed that the epidemiology of visceral leishmaniasis there was different, and very little was known about the reservoirs, vectors, and other aspects of the disease.

Work in Peru showed that good protection could be obtained against *uta*, one of the forms of American tegumentary leishmaniasis, by applying DDT in dwellings and animal sheds and on nearby surfaces, such as stone walls and tree trunks. In Peru, *Phlebotomus* species spread both *uta* and Peruvian warts (*verruga peruana*, a phase of Carrion's disease, or bartonellosis). The incidence of both had fallen, but control of tegumentary leishmaniasis among populations living in forested areas was still unsolved. In Panama, separating dwellings from the forest with a cleared zone was seen to provide a certain degree of protection. When Hertig visited Costa Rica with the entomologist Alexander Graham Bell Fairchild, at the invitation of that country's government and the International Cooperation

Administration (ICA), they suggested experimental use of DDT in forests.[20]

It should be said in passing that a similar practical challenge occurred with bromeliad or forest malaria, first described by Adolpho Lutz (1903, pp. 282–292) and in the 1950s studied by Garnham and Deane (who collaborated closely) in Brazil and Africa. In 1954, this form of malaria sickened approximately four hundred people in Trinidad, the largest and most populated of the 23 islands that comprise Trinidad and Tobago, a British colony that would only gain its independence in 1962. The program to eradicate this disease focused on *Anopheles bellator*, which reproduces in the water that accumulates in bromeliad plants. Employees of the Trinidad and Tobago Health Department tried to eradicate bromeliads with copper sulfate (Rude, 1954). In certain forest areas of Brazil like Santa Catarina, insecticides, larvicides, and chemical substances were also sprayed to destroy bromeliads, resulting in severe environmental impacts that would only be acknowledged afterward (Thiago, 2003, p. 43).

According to the specialists involved in *Tropical Health: A Report*, the two essential aspects of all forms of leishmaniasis remained partially or entirely unsolved: the reservoirs of *Leishmania* and the means of transmission for the diseases it caused in man. They had evidence that all the leishmaniases were transmitted by phlebotomine sandflies, but not necessarily only these insects. In Kenya, for example, the role of termites was investigated.[21] Deane felt an urgent need to determine the vector species of *Phlebotomus* in each region and study their bionomics: domesticity, longevity, seasonal variation, food preferences, and so on.

Another fundamental problem that remained unsolved involved the natural hosts of *Leishmania*, starting with humans. The so-called aberrant clinical forms manifested long after an apparent cure in cases of cutaneous or visceral leishmaniasis, and it was therefore crucial to know whether these patients were still infectious. The diagnostic methods did not address the problem and often were not specific with cross-reactions occurring with Chagas disease and other infections.

[20] *The International Cooperation Administration (ICA) was* the US agency responsible for external assistance and nonmilitary security programs from June 30, 1955, to September 4, 1961. It was the predecessor of the US Agency for International Development (USAID).

[21] Garnham considered mechanical transmission by ticks and stable flies possible, as well as contagion via the respiratory or digestive tract and (more rarely) congenital or venereal infection.

Studies on the reservoirs had to be extended to wild animals. For visceral leishmaniasis, Deane had incriminated wild foxes, and Soviet researchers, the jackals. Ronald B. Heisch then found *Leishmania* in rodents in Kenya. In the Mediterranean region and in China, infected domestic dogs were considered the main reservoirs. Meanwhile in India, for reasons that were not yet explained, dogs were not infected, and humans were assumed to be the only reservoir there.

As for cutaneous leishmaniasis in the Old World, dogs were found to be sensitive to inoculation with *Leishmania tropica minor*, the agent of the classical (dry) form in cities, but this did not affect transmission of the rural (wet) disease. Infected desert rodents, especially the gerbil (*Rhombomys opimus*), had been found in Turkmenistan,[22] Iran, and in Soviet Central Asia. For Garnham and the other participants in this discussion, further study of Old World cutaneous leishmaniasis as a zoonosis was required.

This issue took on more importance in New World cutaneous and mucocutaneous leishmaniasis, which was almost always associated with forests. According to many Latin American researchers, dogs were rarely infected by *L. braziliensis*.[23] The most promising results were obtained by Aristides Herrer (1948, 1949–1951) of the Lima National Institute of Hygiene and Public Health. In an area of Peru where *uta* was raging, 40% of the 469 dogs Herrer examined were infected.[24] As for wild animals, since 1922 Émile Brumpt had included a photo of an agouti (*Dosyprocta azarae*) in the third edition of his *Précis de parasitology,* which was first

[22] In Turkmenistan, a country in Central Asia, outbreaks of human infections of *L. tropica* were found to be associated with infected wild gerbils and ground squirrels. *Phlebotomus caucasicus* and *P. papatasii* living in burrows of the rodents maintained the infection among them. *Phlebotomus papatasii* seemed to be the vector of the parasite from the rodent to man, as observed by Latyshev, Kryukova, and Povalishina (1951), in an article reviewed by the British parasitologist Cecil Arthur Hoare (1954) and by Shekhanov and Suvorova (1960).

[23] Inconclusive results were obtained by Pedroso (1913, 1923) and Leônidas and Maria Deane (1955b) in Brazil, by Mazza (1926, 1927) and Romaña and collaborators (1949) in Argentina; and by Felix Pifano (1940) in Venezuela. Negative results were more frequent in the literature. A good analysis of this literature can be found in Forattini (1960) and Pessôa (1961). The role of dogs in visceral leishmaniasis was analyzed in two substantial theses: Joaquim Eduardo Alencar (1959) [in Portuguese: Canine kalazar: a contribution to the study of its epidemiology in Brazil] and Zigman Brener (1957) [in Portuguese: Canine kalazar in Minas Gerais].

[24] In 1951, the Peruvian physician published four articles on *uta*: they addressed dogs from endemic areas, experimental infection of foxes, the epidemiology of *uta*, and the role of phlebotomine sandflies in the transmission of this cutaneous leishmaniasis (Herrer, 1951a, b, c, d).

published in 1910. He had seen the animal in an endemic zone of São Paulo with ulcers identical to leishmaniasis, but was unable to find parasites in the agouti. Pedro Weiss, of the Lima Institute of Hygiene and Public Health, examined 750 animals in regions of Peru where *espúndia* was epidemic, without success. Pessôa and Barretto (1948) were also unsuccessful in their tests in agoutis, nutrias (*rato-do-banhado, Myocastor coypus*), monkeys, wild pigs, and so on. Oswaldo Paulo Forattini and his colleagues at the School of Hygiene and Public Health of the University of São Paulo examined 928 specimens of forest animals and obtained positive blood cultures for *Leishmania* in only three: a wild rodent (*Kannabateomys amblyonyx amblyonyx*), a paca (*Cuniculus paca paca*), and an agouti (*Dasyprocta azarae*) (Forattini, 1960). In Panama, the wild reservoirs of *Leishmania* remained equally uncertain.[25]

This subject was closely related to another issue, which Leônidas Deane posed in his opinion paper entitled "What is the systematic position of the different *Leishmania* in man?" (NAS, 1962a, p. 505).

Leishmania donovani (Laveran & Mesnil, 1904), *L. tropica* (Wright, 1904), and *L. braziliensis* (Gaspar Vianna, 1911) denoted the parasites for human and canine visceral leishmaniasis, cutaneous leishmaniasis, and American tegumentary leishmaniasis, respectively. Specific or subspecific names had been assigned to *Leishmania* isolated in certain regions, particularly in Latin America, threatening in this case the sovereignty of *L. braziliensis.*

This tendency to associate singular manifestations of leishmaniasis with individual species of *Leishmania* was opposed by the defenders of "Leishmanial unity," namely the notion that all the leishmaniases were caused by a single species, with behavior and clinical manifestations varying according to the local environment and the species of animal or insect that acted as the parasite's host.

None of the attempts to characterize the species based on attributes such as morphology, behavior in culture, infectiveness in laboratory animals, or immunological properties had solved this problem. Even the location in the body was relative, since skin parasitism had been observed in certain cases of visceral leishmaniasis. Visceralization of *L. tropica* and *L. braziliensis* could also occur, but only in laboratory animals, argued Samuel Pessôa (1961, p. 41), a "pluralist." For this parasitologist, the Deanes, Garnham, Adler, and others, each form of leishmaniasis had its own specific agent.

[25] The findings about wild hosts in Panama are analyzed in Lainson and Shaw (1973).

In 1953, Francisco Biagi Filizola of the Universidad Nacional Autónoma de Mexico proposed differentiating the four varieties of American tegumentary leishmaniasis into the mucocutaneous form found in tropical humid forests (*espúndia*), the dry cutaneous form (*uta*), a benign form called *pian-bois*, and *chiclero* ulcer.[26]

The mucocutaneous disease (*espúndia*), which had a significant presence in Brazil and other countries, especially those that shared the Amazon rainforest, erupted in epidemics when forests were cleared to build villages or farms, to construct railroads or highways, or when people moved through these environments during wars or revolutions. In this form of leishmaniasis, nasal complications occurred in approximately 80% of cases, and approximately 30% had lesions in the mucous membranes, leading to repulsive deformities and often death from septic bronchopneumonia.

The Peruvian Pedro Weiss (1943) demonstrated that *espúndia* was propagated in the same areas as malaria, while *uta* raged in high regions of the Andes devoid of forests. This rural disease could spread to urban areas, affected more children than adults, and dogs were frequently parasitized, as in Mediterranean or infantile visceral leishmaniasis.

Pian-bois (French) or *bosch yaws* (Dutch), a benign form of cutaneous leishmaniasis, was found in northern South America and Central America (the Guyanas, Panama, Costa Rica). Small ulcerated lesions healed spontaneously, and the mucosae were invaded in only 5% of cases. *Chiclero* ulcers in Mexico (Yucatan) and Guatemala or *bay sore* in British Honduras only affected the ears and could cause progressive disfigurement of this organ.[27] Diffuse tegumentary leishmaniasis, very similar to lepromatous leprosy, was separately described in 1948 by two Venezuelan dermatologists, Jacinto Convit of the Medical School of the Central University of Venezuela and Pedro Lapenta (who would later direct a leprosarium on Providencia Island, from 1949 to 1953), and by Luiz Prado Barrientos, a professor of tropical diseases at San Andrés University in La Paz, Bolivia.[28]

[26] Biagi (1953) was the Head of the Department of Microbiology and Parasitology at the UNAM medical school. He was the first Mexican doctor to work in the World Health Organization (WHO). He was the author of *Enfermedades parasitarias*, a classic work of Mexican parasitology.

[27] The epidemiology of this form of leishmaniasis was studied by Biagi and by Garnham and Lewis (1959).

[28] Convit and Lapenta (1948), Barrientos (1948). Convit (1958) considered it a new disease and described its characteristics. He adapted a vaccine against leprosy for leishmaniasis. The Venezuelans Medina and Romero (1959) also published on diffuse leishmaniasis.

Diffuse tegumentary leishmaniasis was characterized by the spread of lesions throughout the body and a marked decrease in immune response. Antimonial drugs that were relatively effective in the other leishmaniases had no effect on this form of the disease.

The epidemiology and clinical manifestations of American tegumentary leishmaniasis therefore differed so much from region to region that it was inevitable to suppose there must be several *Leishmania* involved. Biagi (1953) linked *chiclero* ulcer to the subspecies *Leishmania tropica mexicana*. Hervé Alexandre Floch (1954a, b), of the Pasteur Institute of French Guyana and Inini, incriminated *L. tropica guyanensis* as the cause of *pian-bois* and *uta* and *L. tropica braziliensis* as the agent of *espúndia*. In a study published in 1961, Samuel Pessôa (1961) proposed another classification that restored the sovereignty of *Leishmania braziliensis* (Table 5.1).

Garnham and the specialists behind *Tropical Health: A Report* deemed comparative studies essential, thus defending Saul Adler's proposal at the sixth International Congress of Tropical Medicine and Malaria held in

Table 5.1 Synopsis of leishmaniasis classification by Samuel Pessôa

1. Visceral leishmaniasis or *kala-azar*
   Leishmania donovani
2. Oriental sore
   a) Classic or dry type—*L. tropica tropica* (= *L. tropica minor*)
   b) Wet type—*L. tropica major*
3. American tegumentary leishmaniasis
   a) Mucocutaneos leishmaniasis, *ferida brava* or *espúndia*
      L. braziliensis braziliensis (Biagi proposed *Leishmania tropica braziliensis*)
   b) Tegumentary leishmaniasis from the Peruvian mountains or *uta*
      L. braziliensis peruviana (Pessôa used the name *peruviana* suggested by Vélez López in 1913)
   c) Tegumentary leishmaniasis of northern South America and Central America, *pian-bois* or *bosch yaws*
      L. braziliensis guyanensis (Pessôa altered the classification proposed by Floch: *Leishmania tropica guyanensis*)
   d) Mexican tegumentary leishmaniasis or *chicleros* ulcer
      L. braziliensis mexicana (altered the classification proposed by Biagi: *Leishmania tropica mexicana*)
   e) Diffuse tegumentary leishmaniasis—*L. braziliensis pifanoi* (species proposed by Medina & Romero, 1959)

Source: Extracted from Pessôa (1961, pp. 48–49)

Lisbon in 1958 to create centers to maintain *Leishmania* strains isolated in different parts of the world. In 1965, this project became a reality and can be seen as an important milestone in WHO involvement in the leishmaniases problem. An agreement between this agency and Hebrew University in Jerusalem created the first WHO International Reference Centre for Leishmaniasis, directed by Adler (World Health Organization, 1965).

An important collaborator for Adler was Marcello de Vasconcellos Coelho. In 1961, when he took over as head of the Leishmaniasis Research Center Laboratory in Belo Horizonte, Adler was in Brazil and visited this institution. Adler was also one of the members of the working group set up by Amilcar Vianna Martins, director of DNERu, to analyze problems related to leishmaniasis and other rural endemics prevalent in Brazil (Brasil, Departamento…, 1962). Marcello Coelho completed a six-month internship at Hebrew University shortly afterward through a Special Research Training Grant from the World Health Organization. There, with Adler, he studied techniques for maintaining samples of *Leishmania*, and especially immunobiological techniques used to decipher the antigenic composition of various strains in order to identify them specifically.

Both Coelho and Adler presented their work in the seventh International Congress of Tropical Medicine and Malaria, held in Rio de Janeiro in September 1963. The professor at the Hadassah Medical School, who would die three years later (on January 26, 1966), presented his efforts to differentiate *L. braziliensis* from *L. mexicana* and *L. tropica* by comparing their behavior when raised on a semisolid medium made from homologous and heterologous sera (Adler, 1963).

In British Honduras, Ralph Lainson of the London School of Hygiene and Tropical Medicine and John Strangways-Dixon, an entomologist from the University of the West Indies in Kingston, Jamaica, had been trying to experimentally transmit the *Leishmania* responsible for *chicleros* ulcer via *Phlebotomus* since 1959. Various wild-captured species were fed from lesions on hamsters, and then refed on human volunteers. During the second blood meal, four days after the first meal on the infected hamster, the phlebotomine fly[29] transmitted the parasite to a volunteer: J.S.D.,

[29] In an article published in 1963, the insect was classified as *Phlebotomus paraensis* Costa Lima, but was later reclassified as *Lutzomyia pessoana*. Subsequent studies carried out in Belize and the Yucatan Peninsula indicated that the vector for *L. mexicana* was *Lutzomyia olmeca*.

Lainson's colleague. The successful experiment was reported in a research note (Strangways-Dixon & Lainson, 1962). "It was the first experimental proof of transmission of a neotropical *Leishmania* to man through the bite of a phlebotomine," writes Shaw (2015, p. 225).

Saul Adler and especially P.C.C. Garnham were key players in constructing a new international network addressing the leishmaniases, linking mostly isolated teams in the Old and New Worlds with the support of WHO, PAHO, and organizations such as the Wellcome Trust. I am studying this network from the viewpoint of the New World leishmaniases, which received more exposure after research conducted by British scientists in Central America and then Brazil. After the 1964 military coup, large-scale public works in the interior of Brazil transformed American tegumentary leishmaniasis into an extremely serious problem in the Amazon region (Benchimol & Peixoto, 2022; Guerra et al., 2015). In 1965 at the Evandro Chagas Institute in Belém, Pará, Ralph Lainson and Jeffrey Shaw, who were Garnham's disciples at the London School of Hygiene and Tropical Medicine, founded the first Wellcome Parasitology Unit. Lainson, Shaw, and their foreign and Brazilian collaborators showed that populations of parasites in the American continent, with their respective vectors and vertebrate hosts, were much more heterogeneous than had been imagined.[30] Philip Davis Marsden, who also came from the London School of Hygiene and Tropical Medicine, began to teach tropical medicine in 1967 at the University of Brasília and soon became a leading authority in controlling the leishmaniases. Jorge Arias and Toby Vincent Barrett established the Leishmaniasis and Chagas Disease Laboratory at the National Institute of Amazonian Research (INPA) in the 1980s.[31]

Research on the leishmaniases was stimulated by the WHO Special Program for Research and Training in Tropical Diseases (TDR), which was conceived in the 1970s to bridge the science gap in the study and control of neglected diseases affecting impoverished populations in the Americas, Africa, and Asia. In the 1980s, experts mobilized by WHO produced inventories on national leishmaniasis control programs, which for

[30] Among many publications, see Lainson (2010, pp. 13–32).

[31] The work done by these researchers and their institutions is analyzed by Benchimol and Peixoto (2022).

the most part were very precarious, and began to publish recommendations on how the participating nation states should work to achieve control of this complex of diseases on multiple fronts. In May 2007, the 60th World Health Assembly adopted a resolution on controlling the leishmaniases, in which it urged member states to make more consistent efforts toward establishing epidemiological surveys, control programs, systems for monitoring and data collection and analysis, actively seek out cases and access to means of diagnosis and treatment. These states were expected to more effectively promote prevention, which included greater awareness of the leishmaniases at the community level as well as cooperating with neighboring countries that shared endemic areas. The resolution also called on WHO to take the lead in establishing effective control programs in the affected regions (World Health Organization, 2007). In future work, I hope to analyze the varied interfaces between the PAHO and WHO control programs and the countries in the Americas that found themselves at odds with the resurgence of both human and canine visceral leishmaniasis as well as American tegumentary leishmaniasis.

Final Considerations

Ralph Lainson died in Belém, on May 5, 2015. On August 11 of that same year, Elizabeth Rangel, a renowned expert on the insect vectors of *Leishmania* who had established working relations with Lainson and Shaw and Leônidas and Maria Deane (who in turn had worked in the Oswaldo Cruz Foundation in the 1980s up to mid-1990s), spoke about leishmaniasis and these researchers at the III Commemorative Meeting of the National Week for the Control of the Leishmaniases at the Oswaldo Cruz Institute/Fiocruz. I was in the audience to familiarize myself with a subject that was new to me; the lectures were excellent and very instructive, but I left deeply impressed by the quantity of problems and controversies still surrounding this complex group of diseases. There could be no better stimulus for a historian interested in science and health.

During the late-twentieth century in Brazil and other countries, all forms of leishmaniasis emerged or reemerged in urban and rural areas due to environmental changes, human migrations, chaotic urban growth, wars, and other socioeconomic processes underway in large expanses of these countries.[32] Visceral leishmaniasis acquired severe forms when occurring concomitantly with other infections, such as AIDS.

The leishmaniases were then classified as neglected tropical diseases. Although they are in fact overlooked by public policies and affect neglected populations, they mobilize one of Brazil's most vigorous research communities. This is partly due to the fact that many uncertainties still remain about its transmission mechanisms, diagnostic techniques, treatment, and prevention.

At the same time, investigations on the leishmaniases were reframed by new paradigms, especially molecular biology, with increasing numbers of students trained in this discipline in graduate courses that spread across Brazil and other countries in the Americas. Molecular biology has had a major impact on knowledge production about all forms of the disease, including the development of diagnostic techniques, therapeutics, and vaccines.

The growing specialization among the professionals who address the leishmaniases and other neglected diseases today would appear to be counterbalanced (with exceptions, of course) by an inability to perceive the problem holistically, unlike the multivalent generation of P.C.C. Garnham, Samuel Pessôa, and Leônidas Deane, who were sensitive to the social and environmental determinants of the leishmaniases and other endemic diseases. On the other hand, ethical standards for scientific research have changed considerably, and, as a result, some practices that used to be routine in human and animal research before the postwar period have become unthinkable among new generations of physicians and researchers. This is because the leishmaniases, like any other disease, are also a cultural event which depends on the categories of thought and verbal constructs that are specific to a generation, reflecting the history of the medical field and the surrounding society.

[32] In Pará, for instance, the situation of visceral leishmaniasis has not changed in the decades following the studies done by Evandro Chagas and his team, "however, in the early 1980s the disease resurfaced with greater frequency in rural foci and in the suburban and urban areas of medium-size cities such as Santarém ... Over the past two decades the process of expansion increased due to deforestation, socioeconomic factors, and unorganized occupation in the outskirts of cities, where the presence of *Lutzomyia longipalpis* in peri-domiciliar human area and domesticated dogs highly susceptible to infection have facilitated its spread. Today, AVL [American visceral leishmaniasis] has already arrived in the Metropolitan Region of Belém (Cotijuba island), capital of Pará." (Silveira et al., 2016, p. 15).

Acknowledgments This research received funding from the Fundação Carlos Chagas Filho de Amparo à Pesquisa do Estado do Rio de Janeiro (FAPERJ) through the Cientista do Nosso Estado program.

References

Adler, S. (1940a). Attempts to Transmit Visceral Leishmaniasis to Man: Remarks on the Histopathology of Leishmaniasis. *Transactions of the Royal Society of Tropical Medicine and Higiene, 33*, 419–437.

Adler, S. (1940b). Notas sobre *Leishmania chagasi. Memórias do Instituto Oswaldo Cruz, 35*, 173–175.

Adler, S. (1963). Differentiation of Leishmania Brasiliensis from *L. mexicana* and *L. tropica*. In *Proceedings of the 7th International Congress on Tropical Medicine and Malaria* (pp. 308–309). OMS. (Vol. 2, Division A: Tropical Medicine: Helminthic and Protozoal Infections).

Adler, S., & Theodor, O. (1925a). The Experimental Transmission of Cutaneous Leishmaniasis to Man from *Phletotomus papatasii. Annals of Tropical Medicine & Parasitology, 19*, 365–371.

Adler, S., & Theodor, O. (1925b). The Experimental Transmission of Cutaneous Leishmaniasis to Man from *Phletotomus papatasii* (Letter to Editor). *Nature, 116*, 314–315.

Alencar, J. E. d. (1959). *Calazar canino: contribuição para o estudo da epidemiologia do calazar no Brasil.* Tese livre-docência, Universidade do Ceará. Imprensa Oficial.

Alencar, J. E. d. (1961). Profilaxia do calazar no Ceará. *Revista do Instituto de Medicina Tropical de São Paulo, 3*, 177–178.

Alencar, J. E. (1962). Investigações em torno de foco de Calazar na Paraíba. *Revista Brasileira de Malariologia & Doenças Tropicais, 14*, 367–369.

Alencar, J. E., Holanda, D., & Cavalcanti, J. D. N. (1956). Calazar no Vale do Jaguaribe, Ceará, 1955. *Revista Brasileira de Malariologia & Doenças Tropicais, 8*, 33–47.

Aragão, H. d. B. R. (1922). Transmissão de leishmaniose no Brazil pelo *Phlebotomus intermedius. Brazil-Medico, 36*, 129–130.

Aragão, H. d. B. R. (1927). Leishmaniose tegumentar e sua transmissão pelos phlebotomus. *Memórias do Instituto Oswaldo Cruz*, 20, 177–186. (English Version, pp. 187–195).

Barbeitas, M. M.. (2022). *La recherche et le développement pharmaceutique pour les maladies négligées sous l'angle de la leishmaniose:* rupture ou continuité dans la santé globale? Thèse Doctorat, École des Hautes Études en Sciences Sociales.

Barreto, D. C. d. S. (2012). *Uma trajetória familiar na ciência:* Evandro Chagas (1905–1940) e o estudo das endemias rurais no Brasil. Dissertação, Fundação Oswaldo Cruz, Casa de Oswaldo Cruz.

Barrientos, L. P. (1948). Um caso atípico de leishmaniose cutaneomucosa (espúndia). *Memórias do Instituto Oswaldo Cruz, 46*, 415–418.

Benchimol, J. L., & Jogas, D. G., Jr. (2020). *Uma história das leishmanioses no Novo Mundo (fins do século XIX aos anos 1960)*. Editora Fiocruz.

Benchimol, J. L., & Peixoto, C. d. O. (2022). *Uma história das leishmanioses no Novo Mundo (anos 1960 ao século XXI) – Amazônia*. Garamond: Editora Fiocruz. [in print].

Benchimol, J. L., Gualandi, F. d. C., Barreto, D. C. d. S., & Pinheiro, L. d. A. (2019). Leishmaniasis: Historical Configuration in Brazil with an Emphasis on the Visceral Disease, from the 1930s to the 1960s. *Boletim do Museu Paraense Emílio Goeldi. Ciências Humanas, 14*, 611–626.

Biagi, F. (1953). Algunos comentários sobre las leishmaniasis y sus agentes etiológicos: Leishmania tropica mexicana, nueva subespécie. *Medicina, 33*, 401–406.

Brahmachari, S. U. (1940). *Gleanings from My Researches, Vol. 1: Kala-azar, Its Chemotherapy*. University of Calcutta.

Brasil. Departamento Nacional de Endemias Rurais. (1962). *Combate às endemias rurais no Brasil*. (Relatórios dos Grupos de Trabalho reunidos em 1960 na cidade do Rio de Janeiro). Ministério da Saúde.

Brener, Z. (1957). *Calazar canino em Minas Gerais*. Tese Doutorado, Universidade de Minas Gerais.

Brown, T., Cueto, M., & Fee, E. (2006). A transição da saúde pública "internacional" para "global" e a Organização Mundial da Saúde. *História, Ciências, Saúde – Manguinhos, 13*, 623–647.

Brumpt, É. (1922). *Précis de parasitologie*. Masson.

Campos, A. L. V. d. (2006). *Políticas internacionais de saúde na Era Vargas: o Serviço Especial de Saúde Pública, 1942–1960*. Editora Fiocruz.

Chagas, E. (1936a). Visceral leishmaniasis in Brazil. *Science, 84*, 397–398.

Chagas, E. (1936b). Primeira verificação em individuo vivo da leishmaniose visceral no Brasil. (Nota prévia). *Brasil-Médico, 11*, 221–222.

Chagas, A. W. (1939). Infecção de *Phlebotomus intermedius* pela *Leishmania chagasi*. *Brasil-Médico, 1*, 1–2.

Chagas, E., da Cunha, A. M., Castro, G. d. O., Ferreira, L. d. C., & Romaña, C. (1937). Leishmaniose visceral americana: nova entidade mórbida do homem na América do Sul. Relatório dos trabalhos realizados pela comissão encarregada do estudo da leishmaniose visceral americana em 1936. *Memórias do Instituto Oswaldo Cruz, 32*, 321–389, and 88 p. with Maps and Figures.

Chung, H. L. (1940). On the Relationship Between Canine and Human Kala-azar in Peiping and the Identity of *Leishmania donavani* and *Leishmania canis*. *The Chinese Medical Journal, 37*, 501–523.

Convit, J. (1958). Leishmaniasis tegumentaria difusa: nueva entidad clínico-patológica y parasitaria. *Revista de Sanidad y Asistencia Social, 23*, 1–28, Fig. 1-49.

Convit, J., & Lapenta, P. (1948). Sobre um caso de leishmaniose tegumentária de forma disseminada. *Revista de La Policlinica, 17*, 153–158.

Cueto, M. (2015). *Saúde global: uma breve história*. Editora Fiocruz.

Deane, L. M. (1958). Epidemiologia e profilaxia do calazar americano. *Revista Brasileira de Malariologia & Doenças Tropicais, 10*, 431–444.

Deane, L. d. M. (1986). História do Instituto Evandro Chagas: período 1936–1949. In *Instituto Evandro Chagas; 50 anos de contribuição às ciências biológicas e à medicina tropical* (Vol. 1, pp. 53–67). Fundação Serviços de Saúde Pública.

Deane, L. M., & Deane, M. P. (1954). Encontro de leishmanias nas vísceras e na pele de uma raposa, em zona endêmica de calazar, nos arredores de Sobral, Ceará. *O Hospital, 65*, 419–421.

Deane, M. v. P., & Deane, L. M. (1955a). Observações sobre a transmissão da leishmaniose visceral no Ceará. *O Hospital, 48*, 95–114.

Deane, M. v. P., & Deane, L. d. M. (1955b). Leishmaniose visceral urbana (no cão e no homem) em Sobral, Ceará. *O Hospital, 47*, 75–87.

Deane, L. M., Deane, M. P., & Alencar, J. E. (1955). Observações sobre o combate ao *Phlebotomus longipalpis* pela dedetização domiciliária, em área endêmica de calazar no Ceará. *Revista Brasileira de Malariologia & Doenças Tropicais, 7*, 131–141.

Dedet, J. P. (2005). Stages in the Identification of Phlebotomine Sandflies as Vectors of Leishmaniases and Other Tropical Diseases. *Parassitologia, 47*, 291–295.

Dedet, J. P. (2007). Histoire de la médecine: les découvertes d'Edmond Sergent sur la transmission vectorielle des agents de certaines maladies infectieuses humaines et animales. *Bulletin de la Société de Payhologie Exotique, 100*, 147–150.

Desowitz, R. S. (1991). *The Malaria Capers: Tales of Parasites and People*. W. W. Norton.

Dutta, A. K. (2008). Pursuit of Medical Knowledge: Charles Donovan (1863–1951) on Kala-azar in India. *Journal of Medical Biography, 16*, 72–76. http://www.ncbi.nlm.nih.gov/pubmed/18463075. Accessed 5 Sept 2019.

Dutta, A. K. (2009). Medical Research and Control of Disease: Kala-azar in British India. In *The Social History of Health and Medical in Colonial India, Biswamoy Pati, and Mark Harrison* (pp. 93–112). Routledge.

Edler, F., & Pires-Alves, F. (2018). A educação médica: do aprendiz ao especialista. In L. A. Teixeira, T. S. Pimenta, & G. Hochman (Eds.), *História da Saúde no Brasil* (pp. 101–144). Hucitec.

Falcão, A. R. (1963). Susceptibilidade ao DDT e ao Dieldrin, de uma população normal de *Phlebotomus longipalpis*, de Minas Gerais, Brasil. *Revista Brasileira de Malariologia & Medicina Tropical, 15*, 411–415.

Ferreira, L. d. C., Deane, L. d. M., Filho, O. M., & Chagas, A. W. (1938). Notas sobre a transmissão da Leishmaniose Visceral Americana. Trabalho do Instituto

Oswaldo Cruz e do Instituto de Pathologia Experimental do Norte. *O Hospital, 14*, 1077–1087.
Floch, H. (1954a). Leishmania tropica guyanensis n. s.p.: agent de la leishmaniose tégumentaire des Guyanes e de l'Amérique Centrale. *Archives de l'Institut Pasteur de la Guyane Française, 15*, 1–4.
Floch, H. (1954b). Leishmania tropica guyanensis n. sp. : agent de la leishmaniose tégumentaire des Guyanes e de l'Amérique Centrale. *Bulletin de la Societé de Pathologie Exotique, 47*, 784–787.
Forattini, O. P. (1960). Sobre os reservatórios naturais da leishmaniose tegumentar americana. *Revista do Instituto de Medicina Tropical de São Paulo, 2*, 195–203.
Garnham, P. C. C. (1960). *Problemes que posent a l'echelle mondialle la leishmaniose, l'amibiase, la trypanosomiases et la toxoplasmose.* WHO/PA/101.60. Genève: Organisation Mondiale de la Santé. [World Problems in Leishmaniasis, Amoebiasis, Trypanosomiasis and Toxoplasmosis. MHO/PA/101.60 – D60.2161. World Health Organization].
Garnham, P. C. C., & Lewis, D. J. (1959). Parasites of British Honduras with Special Reference to Leishmaniasis. *Transactions of the Royal Society of Tropical Medicine and Hygiene, 53*, 12–35.
Gualandi, F. d. C. (2013). *Medicina Tropical No Brasil:* Evandro Chagas e os estudos sobre a Leishmaniose visceral americana na década de 1930. Dissertação, Fundação Oswaldo Cruz, Casa de Oswaldo Cruz.
Guerra, J. A. d. O., et al. (2015). Tegumentary Leishmaniasis in the State of Amazonas: What Have We Learned and What Do We Need? *Revista da Sociedade Brasileira de Medicina Tropical, 48*, 12–19. Supl 1.
Herrer, A. (1948). Nota preliminar sobre leishmaniosis natural en perros. *Revista de Medicina Experimental, 7*, 62–69.
Herrer, A. (1949–1951). Estudios sobre leishmaniasis tegumentaria en el Peru. *Revista de Medicina Experimental, 8*, 9–137.
Herrer, A. (1951a). Estudios sobre leishmaniasis tegumentária en el Peru, VI: relación entre leishmaniasis tegumentaria y Phlebotomus. *Revista de Medicina Experimental, 8*, 119–137.
Herrer, A. (1951b). Estudios sobre leishmaniasis tegumentária em el Peru, V: leishmaniasis natural em perros procedentes de localidades utógenas. *Revista de Medicina Experimental, 8*, 87–117.
Herrer, A. (1951c). Estudios sobre leishmaniasis tegumentária em el Peru, IV: observaciones epidemiológicas sobre la uta. *Revista de Medicina Experimental, 8*, 45–86.
Herrer, A. (1951d). Estudios sobre leishmaniasis tegumentária en el Peru, II: infección experimental de zorros com cultivos de leishmanias aisladas de casos de uta. *Revista de Medicina Experimental, 8*, 29–33.

Heyneman, D., Hoostraal, H., & Djigounian, A. (1980). *Bibliography of Leishmania and Leishmanial Diseases*. Special Publication. United States Naval Medical Research Unit Number Three (NAMRU-s), Cayro, Egypt. 2 v.

Hoare, A. (1954). Summary of Recent Abstracts. *Tropical Diseases Bulletin, 51*, 37–40.

Hochman, G. (2009). 'O Brasil não é só doença': o programa de saúde pública de Juscelino Kubitschek. *História, Ciências, Saúde – Manguinhos, 16*, 313–331. Supl.

Kemp, A., & Edler, F. (2004). Medical Reform in Brazil and the US: A Comparison of Two Rhetorics. *História, Ciências, Saúde – Manguinhos, 11*, 569–585.

Killick-Kendrick, R. (2013). The Race to Discover the Insect Vector of Kala-azar: A Great Saga of Tropical Medicine 1903–1942. *Bulletin de la Société de Pathologie Exotique, 106*, 131–137.

Lainson, R. (2010). Espécies neotropicais de Leishmania: uma breve revisão histórica sobre sua descoberta, ecologia e taxonomia. *Revista Pan-Amazônica de Saúde, 1*, 13–32.

Lainson, R., & Shaw, J. J. (1973). Leishmanias and Leishmaniasis of the New World, with Particular Reference to Brazil. *Paho Bulletin, 7*, 1–19.

Latyshev, N. I., Kriukova, A. P., & Povalishina, T. P. (1951). Essays on the Regional Parasitology of Middle Asia, I: Leishmaniosis in Tadjikistan. Materials for the Medical Geography of Tadjik S.S.R (Results of Expedition in 1945–1947). *Problems of Regional, General & Experimental Parasitology & Medical Zoology, 7*, 35–62. (em russo). Abstract: *Tropical Diseases Bulletin* 51. 1954. http://citeseerx.ist.psu.edu/viewdoc/download?doi=10.1.1.941.8641&rep=rep1&type=pdf. Accessed 25 May 2019.

Laveran, A., & Mesnil, F. (1904). Sur un Protozoaire nouveau (Piroplasma Donovani Lav. et Mesn.), parasite d'une fièvre de l'Inde. Séance du 7 décembre 1903. *Comptes Rendus Hebdomadaires des Séances de l'Académie des Sciences, 137*(23), 957–961.

Lima, N. T. (2002). O Brasil e a Organização Pan-Americana da Saúde: uma história em três dimensões. In *Caminhos da saúde no Brasil* [online], ed. J. Finkelman, 24–116. Editora Fiocruz.

Lopes, G. (2019). Anopheles Gambiae in Brazil: The Background to a "Silent Spread", 1930–1932. *História, Ciências, Saúde – Manguinhos, 26*, 1–16.

Lu, C. C., Chung, H. L., Ling, C. C., Wu, C. C., Wang, C. C., Chiang, Y. T., & Hsu, W. F. (1955). New China's Achievements in the Treatment and Prevention of Kala-azar. Being One of the Papers Presented by the Medical Delegation of the Chinese People's Republic [...] to the Scientific Session Held Jointly by the Academy of Medical Science of the U. S. S. R. and the Ministry of Public Health of the Uzbee S.S.R. on Problems of Regional Pathology from September 20 to 25, 1954 at Tashkent. *Chinese Medical Journal, 73*, 91–99.

Lutz, A. (1903). Waldmosquitos und Waldmalaria. *Centralblatt für Bakteriologie, Parasitekunde und Infektionskrankheiten, 33*, 282–292.

Lutz, A., & Neiva, A. (1912). Contribuição para o conhecimento das espécies do gênero Phlebotomus existentes no Brasil / Zur Kenntnis der brasilianischen Phlebotomusarten. *Memórias do Instituto Oswaldo Cruz, 4*(1), 84–95.

Marsden, P. D. (1985). Pentavalent Antimonials: Old Drugs for New Diseases. *Revista da Sociedade Brasileira de Medicina Tropical, 18*, 187–198.

Mazza, S. (1926). Leishmaniosis tegumentária y visceral. *Boletin del Instituto de Clínica Quirúrgica, 2*, 209–216.

Mazza, S. (1927). Leishmaniosis cutánea en el caballo y nueva observación de la misma en el perro. *Boletin del Instituto de Clínica Quirúrgica, 3*, 462–464.

Medina, R., & Romero, J. (1959). Estudio clínico y parasitológico de una nueva cepa de Leishmania. *Archivos Venezoelanos de Patología Tropical y Parasitología Medica, 3*, 298–326.

Migone [Mieres], L. E. (1913). Un cas de kala-azar a Assunción (Paraguay). *Bulletin de la Société de Pathologie Exotique, 6*, 118–120.

Myrdal, G. (1952). Les aspects économiques de la santé. *Chronique de l'Organisation Mondiale de la Santé, 6*, 224–242.

National Academy of Sciences-National Research Council (U.S.). Division of Medical Sciences. (1962a). *Tropical Health. A Report on a Study of Needs and Resources.* National Academy of Sciences-National Research Council. Publications 996.

National Academy of Sciences-National Research Council (U.S.). Division of Medical Sciences. (1962b). *Tropical Health. Summary Report on a Study of Needs and Resources.* National Academy of Sciences-National Research Council.

Paraense, L., & Chagas, A. W. (1940). Transmissão experimental da Leishmaniose visceral americana pelo *Phlebotomus intermedius.* Nota Prévia. *Brasil Médico, 12*, 179–180.

Pedroso, A. M. (1913). Leishmaniose local do cão. *Annaes Paulistas de Medicina e Cirurgia, 1*, 33–39.

Pedroso, A. M. (1923). Infecção do cão pela Leishmania tropica. *Revista de Medicina, 4*, 42–44.

Penna, H. d. A. (1934). Leishmaniose visceral no Brasil. *O Brazil-Medico, 48*, 949–952.

Pessôa, S. B. (1961). Classificação das leishmanioses e das espécies do gênero Leishmania. *Arquivos de Higiene e Saúde Pública, 26*, 41–50.

Pessôa, S. B., & Barretto, M. P. (1948). *Leishmaniose Tegumentar Americana.* Ministério da Educação e Saúde, Serviço de Documentação, Imprensa Nacional.

Pifano, F. (1940). La leishmaniasis tegumentaria en el estado Yaracuy, Venezuela. *Revista de la Policlínica, 9*, 3639–3658.

Rangel, E. F., & Shaw, J. J. (Eds.). (2018). *Brazilian Sand Flies.* Springer International Publishing AG.

Romaña, C., et al. (1949). Leishmaniosis tegumentária em perros de Tucumán, II: foco domestico de leishmaniosis. *Anales del Instituto de Medicina Regional, 2*, 283–292.

Rude, M. (1954). *Malaria Eradication and Bromeliad Control in Trinidad.* WHO/5822. WHO Archives, Fonds Photos.

Sergent, E., Sergent, E., Parrot, L. M., Donatien, A. L., & Beguet, M.-E. (1921). Transmission du clou de Biskra par le Phlebotome (*Phlebotomus papatasi* Scop.). *Comptes Rendus Hebdomanires des Séances et Mémoires de l'Academie des Sciences de Paris, 173*, 1031–1032.

Sergent, E., Sergent, E., Parrot, L. M., Donatien, A. L., & Beguet, M.-E. (1926). Transmission expérimentale du bouton d'orient (clou de Biskra) à l'homme par *Phlebotomus papatasi* (Scopoli). *Annales de l'Institut Pasteur de Paris, 40*, 411–430.

Shaw, J. J. (2015). In Memory of Ralph Lainson: A Parasitologist of Amazing Abilities. *Acta Tropica, 150*, 224–226.

Shekhanov, M. V., & Suvorova, L. G. (1960). Nature Foci of Cutaneous Leishmaniasis in the Southwest of Turkmenistan. (Em russo; sumário em inglês). *Meditsiskaya Parazitologya I Parazitarnye Bolezni, 29*, 524–528. [Referência em Heyneman, Hoostraal, and Djigounian. 1980. 2: 220].

Shortt, H. E. (1967). Saul Adler 1895–1966. Elected F. R. S. 1957. *Biographical Memoirs of Fellows of the Royal Society, 13*, 1–34.

Silveira, F. T., et al. (2016). Revendo a trajetória da leishmaniose visceral americana na Amazônia, Brasil: de Evandro Chagas aos dias atuais. *Revista Pan-Amazônica de Saúde, 7*, 15–22. N. especial.

Soper, F. L., Penna, H., Cardoso, E., Serafim, J., Jr., Frobisher, M., Jr., & Pinheiro, J. F. (1933). Yellow Fever Without *Aedes aegypti:* Study of a Rural Epidemic in the Valle do Chanaan, Espírito Santo, 1932. *The American Journal of Hygiene, 18*, 555–587.

Strangways-Dixon, J., & Lainson, R. (1962). Dermal Leishmaniasis in British Honduras: Transmission of *L. brasiliensis* by *Phlebotomus* Species. *British Medical Journal, 1*, 297–299.

Swaminath, C. S., Shortt, H. E., & Anderson, L. A. P. (1942). Transmission of Indian Kala-azar to Man by the Bites of *Phlebotomus argentipes. Indian Journal of Medical Research, 30*, 473–477.

Telkes, E. (1998). Dictionnaire biographique de la première génération de professeurs de l'Université Hébraïque de Jérusalem, p. 39–51. *Bulletin du Centre de Recherche Français à Jérusalem, 2*, 1–9. https://journals.openedition.org/bcrfj/4512. Accessed 23 May 2018.

Théodoridès, J. (1997). Note historique sur la découverte de la transmission de la leishmaniose cutanée par les phlébotomes. *Bulletin de la Societé de Pathologie Exotique, 90*, 177–178.

Thiago, P. d. T. S. (2003). *História da malária em Santa Catarina.* Dissertação, Universidade Federal de Santa Catarina.

Vélez López, L. R. (1913). Uta et espúndia. *Bulletin de La Societé de Pathologie Éxotique, 6*, 545.

Vianna, G. (1911). Sobre uma nova espécie de Leishmania (Nota preliminar). *Brazil Médico, 25*, 411.

Vianna, G. (1912). Tratamento da leishmaniose pelo tártaro emético. Fala reproduzida na Quarta sessão ordinária da Sociedade Brasileira de Dermatologia realizada em 24 e 25 de abril de 1912, em Belo Horizonte, sob a presidência do Prof Fernando Terra. Em Annaes do VII Congresso Brasileiro de Medina e Cirurgia (Belo Horizonte, Minas Gerais). *Archivos Brasileiros de Medicina, 2*, 422–436.

Vianna, G. (1914). Parasitismo da célula muscular lisa pela *Leishmania braziliensis*. *Memórias do Instituto Oswaldo Cruz, 6*, 39–42.

Weiss, P. (1943). Epidemiologia e clínica de la leishmaniosis tegumentária en el Peru. *Revista de Medicina Experimental, 2*, 209–248.

Winslow, C. E. (1955). *Lo que cuesta la enfermedad y lo que vale la salud*. Organización Mundial de la Salud, Oficina Sanitaria Panamericana.

World Health Organization. (2007). *Resolutions and Decisions*. WHA60.13: Control of Leishmaniasis. Sixtieth World Health Assembly. Ninth Plenary Meeting: Committee A, Second Report. https://www.who.int/neglected_diseases/mediacentre/WHA_60.13_Eng.pdf. Accessed 26 Nov 2018.

World Health Organization. (2010). *Working to Overcome the Global Impact of Neglected Tropical Diseases*. First Report on Neglected Tropical Diseases. http://apps.who.int/iris/bitstream/10665/44440/1/9789241564090_eng.pdf. Accessed 5 Sept 2021.

World Health Organization. (2017). *Leishmaniasis: Fact Sheet*. http://www.who.int/mediacentre/factsheets/fs375/en/. Accessed 10 July 2017.

World Health Organization. Research Programme on Leishmaniasis. (1965). *Service to Research Workers*. WHO/LEISH/1.65 - D65-454 (WHO International Reference Centre for Leishmaniasis, Jerusalem).

Wright, J. H. (1904). Protozoa in a case of tropical ulcer ('Aleppo boil'). *The Journal of Cutaneous Diseases Including Syphilis, 22*(1), 1–9.

Yakimoff, W. L., & Schockor, N. F. (1915). Leishmaniose cutanée (Bouton d'Orient) au Turkestan Russe. Reunion Biologique de Petrograd. Séance de janvier 1915. *Comptes Rendus Hebdomanaires des Séances et Mémoires de la Société de Biologie, 67*, 107–109. https://ia800500.us.archive.org/6/items/comptesrendusdes78soci/comptesrendusdes78soci.pdf. Accessed 21 Aug 2018.

PART III

Circulation of People, Object, and Ideas

CHAPTER 6

Tropical Medicine, the Nation, and the Colonial Expansion in the View of Italian Royal Navy Physicians at the End of the Nineteenth Century

Mauro Capocci and Daniele Cozzoli

Colonialism, Emigration, Biomedicine, and the Making of the Italian Nation

The aim of this chapter is to explore the connections between two contemporary historical phenomena: the Italian colonial expansion towards East and North Africa and the Italian emigration towards the Americas.

M. Capocci (✉)
Department of Civilizations and Forms of Knowledge, University of Pisa, Pisa, Italy
e-mail: mauro.capocci@unipi.it

D. Cozzoli
Department of Humanities, Pompeu Fabra University, Barcelona, Spain
e-mail: daniele.cozzoli@upf.edu

M. Capocci, D. Cozzoli (eds.), *Empire, Nation-building, and the Age of Tropical Medicine, 1885–1960*, Medicine and Biomedical Sciences in Modern History,
https://doi.org/10.1007/978-3-031-38805-7_6

The Italian colonial venture was limited in time and space. Its birth date is usually considered the acquisition of Assab Bay in 1882 and its end date the defeat in Second World War. The Italian protectorate over Somalia lasted until 1960, though, as part of the long process of the demise of the European colonial empires (on Italian colonialism, see Del Boca, 1976–1984; Labanca, 2002). At the same time, starting in the last decades of the nineteenth century, millions of Italians emigrated mostly towards the Americas. The contemporaneity of these two processes makes the short-lived Italian colonialism an interesting case for studying the relationship between modern colonialism and emigration. We will focus on colonialism and emigration during the Age of Empire, as Eric Hobsbawm (1987) defined the period between 1875 and 1914, roughly between the Long Depression and the beginning of the First World War.

It is worth noting that when dealing with the colonialism-emigration nexus, even the language is ambiguous: the word *colonia* in Italian, as in other Romance languages, has a wide semantic field. Firstly, *colonia* means colony, the occupation of lands outside the motherland, which are ruled on behalf of or from the motherland. The Italians, however, also referred to the communities of immigrants in other sovereign states as *colonie*. This was the case of the so-called 'settlement colonies' in South America: groups of immigrants, mostly peasants, who occupied an expanse of land and built up a small village. It is worth noting that the term *colonia* was not only associated with agriculture and land occupation. In 1871, the Italian Society for Political Economy advertized an essay prize about the 'formation of spontaneous colonies of Italians abroad' (quoted in Carpi, 1874, I, p. 11)—just like cells on a Petri dish. Italian immigrants in Switzerland were grouped in the *Federazione delle Colonie Libere* (Federation of the Free Colonies) until the 1960s (Ricciardi, 2013). *Colonia* is also used in the expression *colonia estiva* (summer camp). During Fascism, *colonie estive* were organized for the children of working-class families in the public sector and often for the children of military personnel, reinforcing the military-like education given to Italian youth under Fascism. The group of Italian Jews who emigrated to São Paulo in the late 1930s because of the racial laws referred to themselves as the 'Colonia Mussolini'—Mussolini Summer Camp—meaning, we are here on holiday, thanks to Mussolini (see Cozzoli, 2016). This expression also echoed the notorious *colonia penale* (penal colony), a correction institute where prisoners could be employed in forced labour, and that were usually isolated and faraway, often on small islands. During Fascism, political dissidents were often confined to such remote establishments. Finally, the term *colono*, following the Roman institution, refers to a peasant who is renting a piece of land. Thus, in Italian documents African peasants working on farms owned by Italians in Italian colonies are called *coloni*. In the 1930s Luigi Talamonti, a civilian physician who worked in Villaggio Duca degli Abruzzi, an agricultural settlement near Mogadishu, called the Somalis *coloni* when he wanted to stress the labour relationship and *indigeni* when he wanted to stress their 'otherness' with regard to the Italian workers and landowners (Talamonti, 1933, p. 123).

This semantic complexity mirrors a deep historical relationship between two phenomena of people in movement: emigration, including the internal sort, and colonialism. The relationship is, of course, not only rooted in language, but in historical events and in the consciousness of the contemporary elites. The 'subalterns,' the poorest strata of Italian society, who migrated to South America and settled in agricultural colonies in Africa, could not establish a coherent connection between the two processes. As Antonio Gramsci argued in 1935 in his Prison Notebook 27, the subaltern is by definition unable to have a systematic, coherent worldview. Their conception of the world is—Gramsci argued—fragmentary, a mechanical juxtaposition of various conceptions of the world and life that have followed one another throughout history.[1] The elites, on the contrary, were fully aware of the connection between colonialism and emigration. In 1908 and 1911, before the Italo-Turkish War that led to the conquest of Libya, the *Istituto Coloniale Italiano* (the Colonial Institute) organized two 'congresses of the Italians abroad,' with delegates from European and North and South American countries, but also from Egypt, Tunisia, and the Eritrean colony (ICI, 1910, 1911). Thus, there were Italians in other independent states, Italians in formal and informal colonies of other European powers, and Italians in Italian colonies. All of them were considered part of the same national community.

However, local communities in South America often perceived the Italian newcomers as a threat and associated their arrival with colonial expansion. On 5 September 1896, a few months after the defeat at Adwa, the Italian consul in São Paulo, Edoardo Compans De Brichanteau, alarmingly reported that the local mob shouted: 'Death to the colony, to Italy, to the Consul, long live Menelik,' to the Italian newcomers, who were

[1] Gramsci, 2007, vol. 3, pp. 2311–7. *Prison Notebook 27* is partially translated in Gramsci, 2021, pp. 129–132.

cheering the Italian consul. (Letter from the Royal Consul in Rio de Janeiro to the Minister of Foreign Affairs in Ministero degli Affari Esteri (henceforth MAE), Documenti Diplomatici, Incidente del Brasile, serie C 1892–1894, p. 196). In the period that followed the fall of the Brazilian Empire, this was just one of the episodes of tensions between the Italians and the local communities.

For its part, the Italian colonial lobby was well aware of the connection between emigration and the colonial venture, and of the Italian peculiarity in this respect. In a report delivered to the annual session of the Institut Colonial International in 1905, senator Luigi Bodio—the long-standing secretary general to the Royal Office of Statistics as well as the head of the Italian Emigration Committee—made a clear distinction between exploitation colonies, in which the export of capital was crucial, and population colonies, where the export of manpower was predominant (see Bodio, 1905; on the International Colonial Institute, see Wagner, 2002; on Bodio, see Bonelli, 2018–19 and Cappiello, 2014). Italian emigration towards the Americas clearly fitted the latter definition.

In his *Globalisation and the Nation in Imperial Germany*, Sebastian Conrad focuses on labour in order to deal with the making of the German nation in the evolving global context. Such a fixed point allows him to study emigration flows, including those related to colonial and imperial dominance patterns. In our research, we decided to adopt a different fixed point, that of biomedicine and in particular the work of Navy physicians.[2] This fixed point allows us to look at the complex relationship between colonialism and emigration, and the creation of the Italian nation in the context of the Age of Empire. In 1903, Alessandro Pasquale gave the introductory lecture to the newly created course in hygiene at the University of Naples—*The aim and importance of naval and colonial hygiene* (on Pasquale, see Bonelli, 2018–9). The lecture focused on the Navy doctor as a crucial figure both in colonial expansion and in emigration: Navy physicians were embarked both on military missions around the globe and on the emigration service steamboats. From this perspective, their work provides a privileged view of the way in which the political and cultural elites understood and connected colonialism and emigration. Our fixed point also allows for the observation of the subaltern angle, since Navy doctors reported on local communities, describing how they

[2] The importance of Navy physicians in the making of tropical medicine in Italy has been stressed by Bonelli (2018–2019).

perceived the colonialists, as well as how the Italian immigrants related to them. Military doctors and medical investigators were, indeed, an elite, albeit a peculiar one. They shared some of the values of the European elites—such as the mission of bringing civilization to the natives—but they added the belief in the progressive power of science, the commitment to save human lives, and their military perspective. In 1903, Filippo Rho, the Navy doctor who directed the *Annali di Medicina Navale e Coloniale* and later held the chair of tropical medicine at the University of Naples, explained his view of tropical medicine, colonialism, and emigration:

> [...] medicine within this work of expansion is one of the primary factors of civilization and progress. Firstly, it prepares the defence against exotic diseases and their prevention, whose importation is eased by the frequency and speed of trans-oceanic communication; secondly, it significantly contributes to promote colonial growth by foreseeing and healing the prevailing diseases in remote countries, above all in the tropics. (Rho, 1903, p. 388)[3]

In recent years, Lucy Ryall (2022), Alessandro Bonvini (2021), and Pamela Ballinger (2020) have separately studied different aspects of Italian migration to and from formal and informal colonies. Focusing on the Genoese merchant community in nineteenth-century Peru, Lucy Ryall has pointed to the 'migrant colonialism' in the making of Italian imperialism. Alessandro Bonvini has investigated the action of Italian 'patriots' in the political and military struggles in South America. Pamela Ballinger has observed that after the Second World War refugees from the lost territories contributed to shape the contemporary Italian nation.

Our aim is to contribute to the ongoing historical reflection on Italian colonialism, emphasizing the global connections in the making of the Italian nation in the nineteenth and twentieth centuries and highlighting the role of the medical elite in this process.

The Navy

A focus on the Navy is crucial to understand the nature of Italian colonial expansion throughout the nineteenth century. In 1814, the Kingdom of Sardinia absorbed the Republic of Genoa. The small monarchy was ruled by the autocratic and aggressive House of Savoy, which ultimately aimed

[3] On Filippo Rho, see Bonelli (2018–2019).

to expand its possessions in the Peninsula. It was the military initiative of the Kingdom of Sardinia that ended up unifying the whole Peninsula in a nation-state in 1861. Since the Napoleonic age, Ligurians had begun migrating to South America (especially towards the Rio de la Plata area), and Genoese merchants had established trade routes with the Americas (Gabriele & Friz, 1973). With the occupation of Genoa, the Piedmontese had also acquired a merchant fleet, a Navy, and a special relationship with the Genoese settlement colony. The Kingdom of Sardinia was contemplating an expansion towards the Americas, and the Genoese community in the Americas could be instrumental to such a goal. Between the 1830s and the 1860s, the Savoyards encouraged commercial penetration and promoted scientific expeditions in the Americas. Piedmontese Navy ships explored the oceans (see Leva, 1936–1960; Gabriele & Friz, 1973; Ryall, 2022), and after the creation of the Kingdom of Italy, at least one Navy ship was permanently anchored in the area of the Rio de la Plata estuary, while as early as 1865, a permanent Italian naval force in South America was anchored in the La Plata port (Leva, 1936–60, vol. I, pp. 32–59 and 247–72, vol. II, pp. 159–90, 210–34 and 265–84; Gabriele & Friz, 1973, pp. 95–102). In 1888, the ships were called back to Italy, but the following year a new naval force was sent to the La Plata port (Gabriele & Friz, 1982, p. 73).

In 1892, the Italian foreign minister explicitly stated to his Brazilian counterpart that the battleships were harboured in La Plata to protect the Italian colonies in South America (MAE, Incidente del Brasile, vol. I, p. 9). The Italian fleet in La Plata intervened in local conflicts to protect Italian immigrants, but also to influence events. They were a military instrument of the Italian penetration in the continent: emigration and military pressure were two sides of the same coin. Counter-Admiral Fausto Leva—who authored a four-volume history of the oceanic campaigns of the Italian Navy for the Historical Office of the Navy—argued that the Italian Royal Navy ships in La Plata were crucial to provide the migrants from the Italian Peninsula with a sense of belonging to the newly created Italian nation (Leva, 1936–60, vol. I, p. 21). It is worth noting that the first two volumes were published in 1936, the year of the proclamation of the Empire after the invasion of Ethiopia, the third one in 1940, and the last one in 1960, another crucial date for Italian colonialism, as that year marked the end of the Italian mandate in Somalia and finally terminated Italian colonialism. Leva's remarks mirror the concerns of the contemporary Italian elites during the colonial expansion.

Contemporarily, Navy ships were sent worldwide in order to carry out scientific explorations, sign commercial agreements, and explore the possibility of establishing settlement colonies or even penal colonies—at least until 1889, when the new penal code suppressed deportation of prisoners (Gabriele & Friz, 1973, pp. 207–38).

It is important to stress that the expansion promoted initially by the Kingdom of Sardinia and then by the Kingdom of Italy towards the Americas by means of settlement colonies was the result of both the spontaneous activity of the migrants and a series of plans and projects (often never fully materialized). On the part of the governments and policymakers, there was no single coherent imperialistic design connecting the armed colonialism in Africa with that of the settlement colonies in the Americas.

Navy Physicians and Emigration

Emigration from Italy to the Americas skyrocketed as a consequence of the agrarian crisis that took place in Europe in the last three decades of the century. The progress of steam navigation allowed for the massive import of wheat from the USA and the Russian Empire, and of rice from South Asia to Europe. Prices fell and for many peasants this meant hunger. In Italy, the crisis spanned from 1880 to 1894, but in 1898 the Spanish-American War caused a halt to the import of wheat from the USA, abruptly worsening the situation. Southern Italian peasants were especially affected by the agrarian crisis, and they started migrating towards the Americas (see Salvadori, 2018, pp. 76–8). Emigration was officially regulated as early as 1888, though the law mostly aimed at protecting emigrants from frauds by private emigration agencies. A new bill on emigration was passed in 1901, finally setting health standards for the emigration steamers, including access to freshwater, quality and quantity of food, light, air, and so on (see Madia, 1909–10). Accordingly, the government created a school of colonial policy in Rome and organized courses in naval hygiene and exotic medicine at the University of Naples in order to train civilian and Navy doctors (Rho, 1903, p. 407). The new law on emigration imposed the presence of a physician of the Royal Navy instead of a civilian one. As an employee of the navigation company, a standard physician was subordinated to the captain of the ship and could hardly resist the pressure exerted on him by the company. As a result, the journey to the Americas often turned out to be a nightmare for lowly migrants (see Madia, 1909–10, p. 418). In 1905, Luigi Bodio estimated that there were 50

Navy officers (medical lieutenants and captains) working on emigration steamboats (Bodio, 1905, p. 253). Navy doctors' duties were not limited to health. They had to provide official statistical reports about the emigrants and acted as official representatives of the Italian government (Rho, 1903, p. 407; Pasquale, 1903, p. 377). Captain Ernesto Madia—a medical doctor in the Royal Navy who published detailed reports on his own service on emigration steamboats—praised the new law: the migrants found in the Navy doctor the protection of the state (Madia, 1909–10, p. 421). Madia also explained that the doctor on the emigration service vessels should prevent the emigration of the sick, while improving the personal hygiene and cleanliness of the migrants: the Anglo-Saxon people despised dirtiness and weakness, so that a tidy migrant had greater chances to get through the border (Madia, 1909–10, pp. 432–3). Madia expressed the feeling of shame of the Italian bourgeois class in front of the mass of desperate, dirty, sick people who were leaving a country unable to assure them a decent life. According to Dr. Alessandro Pasquale, a hygiene lecturer at the University of Naples, Navy doctors should select migrants not only by rejecting the sick and unvaccinated, but also by acting as a sort of border police, by stopping those 'deprived of the means necessary to sustain the more urgent needs of their existence' (Pasquale, 1903, p. 377). This view reveals how complicated the relationship was between the Italian liberal state, its military bodies, and the subaltern. The Navy doctor acted as a benevolent guide or as a scrupulous gatekeeper who could end their dreams. In Chap. 1 of this volume, Isabel Amaral has stressed the role of Navy physicians in establishing and developing tropical medicine in Portugal, also connecting the country to the wider network of European colonial researchers. Italian Navy doctors played a similar part, though tropical and colonial medicine went through a different history, lacking, for example, a centralized institution such as the Lisbon School of Tropical Medicine. In following paragraphs, we will analyse the different roles played by the Navy physicians in Italian colonialism.

Navy Physicians as Observers

The body of Navy physicians was one of the few groups that circulated knowledge about the colonies and emigration. On every ship there was at least one doctor, but more often there were two. Alessandro Pasquale recalled the importance of acquiring information:

> The success of colonial expansions is 3/4 entrusted to the exact knowledge of the physical geography of the places one goes to colonize, the climate, fauna, flora, food resources that there are, the diseases that prevail there and the way of fighting against them. (Pasquale, 1903, p. 383)

A primary concern of the Navy doctors was to provide information about the territories they visited, the climate conditions, and the possibility for 'whites' to settle. In November 1894, the Royal Navy battleship *Cristoforo Colombo* set sail for a two-year circumnavigation of the world (see Leva, 1936–60, vol. II, pp. 301–22). The first on-board physician, Carlo M. Belli, often commented on the climate of the North and South American areas where the *Colombo* anchored, duly noting whether the conditions were suitable for European settlements (Belli, 1897, p. 1269). Another Navy physician on board the ship *Umbria*, Dr. Salvatore Accurso, reported in 1898 about the climatic and hygienic conditions of the Somali region of Benadir, at the time an Italian concession from the Sultanate of Zanzibar. His emphasis was on the suitability of the area for European colonies (Accurso, 1898).

According to Patrick Manson's foundational textbook (1898), climate could only be listed as an indirect cause of disease, germs being the agents of most infectious pathologies. However, human acclimatization was a concern for several physicians. The Frenchman Pierre-Just Navarre, Navy physician and author of *Manuel d'hygiène colonial: guide de l'Européen dans les pays chauds* (1895), remarked that 'la race caucasique n'est pas acclimatée aux pays intertropicaux, ... on ne constate aucune adaptation physiologique de l'organisme au nouveau milieu, et que toutes les modifications fonctionnelles aboutissent à un état morbide,' so that the European needed hygiene and medicine for the 'unequal struggle' for life in the tropics (Navarre, 1895, pp. 46–7).

Manson himself pointed out that although the environment was not a direct pathogen, it caused modifications in the human response, so that 'natives of tropical countries... are attuned by heredity and habit to the conditions they were born into' (Manson, 1898, p. XII). This view was echoed in the writings of Filippo Rho (1903, pp. 389–93), who remarked that climate modified human response to environment and disease. Medical knowledge was thus instrumental to a healthy settlement of individuals of European descent in the warm regions.

As Italian Royal Navy medical officers generally agreed with Manson's view, they also shared the concern with acclimatization and environment, and their papers often reported on climate from a medical perspective. Navy lieutenant colonel Umberto Colorni reported in 1900 that he had to stop the irrigation of the camp, since the excess humidity of the soil caused intestinal disorders in the soldiers (Colorni, 1900, p. 277). Likewise, Belli observed that the climate in Los Angeles and San Diego was especially favourable to patients affected by nervous diseases and to those recovering after weakening illnesses (Belli, 1897, p. 1274). A couple of years later, P. Guerra, a first-class physician sailing towards Northern China on the steamship *San Gottardo* in July and August of 1899—some months before the Boxer rebellion—was relieved that the bad climate of Hong Kong did not affect the crew of the ship (Guerra, 1900, p. 646).

Some Italian physicians blamed other factors for susceptibility to tropical disease, including alcohol abuse. Experiencing a yellow fever outbreak in 1895 in El Salvador, Belli argued that Europeans were more affected because they drank too much (Belli, 1897, p. 1278; on the frequent considerations by Italian physicians on alcohol abuse, see Bonelli, 2018–9). Mario Mannelli, second-class physician on board the *Staffetta* (a Royal Navy ship on a mission in the Indian Ocean), also blamed alcohol abuse and spicy food for the gastroenteric diseases common among the Seychelles native populations (Mannelli, 1899, pp. 499–500).

Italian Navy doctors also acted as prolific sources of intelligence about the countries they visited. They were particularly interested in demographics, public health, prevalent diseases, and the state of immigration. They often offered thorough reports about health organizations, medical schools, medical training, and higher education (Belli pp. 1271–3, Nota, 1907). Hospitals of several countries were described: Belli compared the poor condition of hospitals in Central and South America with the excellence he found in North America; Bottini reported on the civilian hospital in Santos and the Boston City Hospital; Roselli on the Roosevelt Hospital; and Luzzati on the training school for nurses at St. Luke's Hospital (Bottini, 1903a, b; Roselli, 1907; Luzzati, 1907). Whenever it was possible, they also gathered knowledge on local flora and fauna, and this was the case of Umberto Colorni, who participated in an expedition in the interior of Morocco (Colorni, 1900). The interests of Italian Navy physicians varied widely, also within the medical subjects. While tropical medicine was possibly their main focus, they were also concerned with all the diseases that could affect sailors and migrants alike in several diverse locations. The 1896 volume of *Annali di Medicina Navale* included articles on beriberi, malaria in Madagascar, and the impact and treatment of heatstroke in the US Navy. The increase of Italian emigration towards the Americas also entailed that Italian Navy doctors had to deal with unusual diseases, often to be treated back in the motherland. Dr. Malizia, a Navy doctor serving on the steamship *Colombo* travelling to Pará and Manaus, visited local hospitals hoping to learn more about yellow fever and beriberi. Indeed, on the journey back to Italy he had to treat passengers affected by beriberi and malaria (Malizia, 1903). Bottini carefully observed the activity of the civilian hospital in Santos as well, noting that in the current year, 338 Italians were treated there (Bottini, 1903a, p. 339). Malaria attracted particular interest—as it was still endemic in large areas of Italy—together with tuberculosis (Guerra, 1900). The attitude by Italian Navy physicians reflected the fact that it was not only a 'colonial' disease: knowledge circulation was much more intense, underlining similarities and differences between the Italian and the colonial malaria. Several analyses of the Italian strategies in the fight against malaria (Fantini, 1998; Snowden, 1999, 2006; Corbellini, 2022) noted a distinct colonial approach towards the disease in the Italian South, further proving the diverse nature of colonialism. In Italy (as well as in the other Southern European countries still affected by malaria), the campaigns targeted its own citizens and also aimed at improving their living and cultural condition, becoming an important piece in the puzzle of Italian nation-making. In tropical settings, such as the Indian and African colonies, this was not always the case, being the disease more intimately connected to 'climate' (Packard, 2007): the main goal of antimalarial campaigns was the protection of Europeans and of the workers they needed, as noted above by DeBarros and Neill in Chap. 3.

Tuberculosis was also deemed an important subject, being common in Europe as well as in most of the American destinations of Italian emigrants. Belli noted that infection was the first cause of death in Boston, as it was in Europe (Belli, 1897, p. 1317). Doctors were also concerned that tuberculosis could be imported from urban to rural areas in Italy (Molinari, 2002). Balduino, the Navy doctor on board the steamship *La Plata* on emigration service from Naples to Buenos Aires in 1903, reported on the struggle against tuberculosis in South America (Balduino, 1903). In order to get information, Navy doctors resorted to a capillary network of Italian doctors working in the Americas. This network allowed Navy doctors to access several hospitals and to meet the Italians treated in those institutes (see Malizia, 1903, p. 363). Some of the hospitals, especially in South America, were created by the local Italian communities to provide healthcare—and pride—for the nationals. Such institutions were duly described by Navy doctors (Belli, 1897, p. 1287; Buonanni, 1903), as a part of their effort to provide an accurate picture of the status of Italian immigrants abroad. Captain Gerardo Olivi provided, for example, a long report on how Italian migrants were treated upon arrival in the USA, stressing the importance of charities providing assistance, such as the *Society for the Protection of the Italian Immigrants*, the *Italian Benevolent Institute* and the *San Raffaele Society* (Olivi, 1903). Furthermore, their reports provide information on habits, urban spaces, and opinions about the people they met, stemming from the curiosity of the observers and appealing to the curiosity of their audience.

Indigenous Knowledge

Navy physicians also observed and commented on indigenous knowledge. Their attitude did not differ much from that of Alexander von Humboldt, who observed indigenous knowledge and practice with curiosity and respect, trying to understand it and elaborate it in terms of Western knowledge and practices. There were, of course, differences and nuances. The *Cristoforo Colombo* mission on the East African coast provides an interesting case. Samples of a poison used by Somali hunters were brought back to Italy: the poison, known as *ouabaio*, is derived from the wood of the *Acokanthera*[4] (Colorni & Rodolfo, 1897; Baroni, 1899, 1900; Gianturco, 1899, 1900). The Italian War Ministry set up a committee to study the *ouabaio* samples, and the Navy appointed a medical doctor, Alessandro Pasquale, and a pharmacologist and chemist, Baroni. The latter studied *ouabaio* together with *curare*, a poison used by Amazon indigenous hunters. Baroni did not specify where the *curare* he used came from.[5] He was not able to isolate the active principles of the three samples, although he correctly understood that *ouabaio* was a glycoside and *curare* an alkaloid (Baroni, 1899, p. 366; see also Baroni, 1900 and Gianturco, 1899, 1900). It is interesting to note that the comparison between the two poisons

[4] Specifically, *Acokanthera schimperi*, which contains ouabain (also known as *g-strophantin*), a toxic compound that in low doses can be used to treat heart diseases. On *g-strophantin*, see Osseo-Asare (2014).

[5] A number of explorers brought back samples of *curare* to Italy in the previous decades. See Moroni & Dell'Acqua, 1863. On the history of curare, see McIntyre, 1947; Thomas, 1964; Smith, 1969; Grmek, 1973; and Cozzoli, 2016. See also Lesch, 1981 and Maehle, 1999.

drew on their function rather than their chemistry, an implicit connection established by Baroni between the worlds of the Amazon and Somali natives.

The focus on indigenous knowledge was also two-faced: attentive when there were hopes of learning something useful, but dismissive in other cases. Colorni, a medical officer on the *Colombo* on a two-year voyage around the world between 1894 and 1896, carefully evaluated the claims of the Somalis in his study of ouabain (Colorni & Rodolfo, 1897) and neutrally reported on a number of indigenous remedies and practices that could be assimilated to the European medical tradition. For instance, he explained that in Tangier the natives used to fight cholera by injecting human pus in the wrist, and that the Moroccans used to stimulate copious sweating in cases of snakebite (Colorni, 1900, p. 274). However, Colorni showed lack of understanding and respect for the Moroccans. The description he made of an encounter with local physicians is revealing:

> One evening at the market of Sid-ben-Nur, four days from Mazagan, I saw two intelligent-looking Arabs in a tent, who I was told were physicians; around the tent were crowding customers, whom, after having heard the account of their own or of some far relative's disease, [the physicians] gave paper sheets on which was written a verse of the Koran. (Colorni, 1900, p. 280)

There is a noticeable rhetorical contrast between the 'two intelligent-looking Arabs' and their lack of any medical knowledge. Similarly evident is that the physicians are not examining the patients. They can prescribe a cure, and a blatantly useless one, and Colorni underlines that the relative is 'far,' leading the reader to think that asking for a treatment for a 'close' relative is better. Colorni commented that, while believing in superstitious remedies, the natives also trusted Western doctors, though only to some extent, as they do not always comply with medical prescriptions (Colorni, 1900, p. 280). The conclusion of Colorni's report is also a product of his bias:

> […] the disease that most concerns him (the Moroccan) is the one I mentioned above, i.e. sexual impotence, patently incompatible with his state of polygamy, but that alas I had to heal with many words and a dose of sodium bicarbonate; perhaps suggestion would replace the ineffective remedy. (Colorni, 1900, p. 281)

Thus, the Western doctor should never admit his limitations to the natives, otherwise he may undermine the natives' belief in the superiority of Western culture. As a result, the European doctor behaves like the Arab physician he criticized earlier, prescribing a useless remedy and adding 'many words' as a replacement for the Quranic verse. The bias against non-Western medicine is also apparent in Guerra's attitude towards a plague-fighting ritual:

> To drive away the evil spirit, who according to the Chinese would be embodied by the plague, they meet up in procession and walk the streets of the Chinese town, setting off fireworks, and pulling long red-hot chains, and they appear to be persuaded and certain, in their blind superstition, that in so doing they can drive away that plague that has instead in their innate dirtiness and their poor nutrition a very powerful aid to its development and rapid spread. (Guerra, 1900, p. 646)

On the other hand, Belli praised the hygienic organisation of burial in Bombay, which respected the Hindu practice of letting the bodies putrefy (Belli, 1897).[6] He also praised Buenos Aires' brothels, for the public displaying of the health status of their prostitutes without implementing unnecessary restrictions on the personal freedom of infected women (Belli, 1897, p. 1298). Belli also highlighted the Salesian missionaries' mistakes with Fuegians in Punta Arenas, since they brought 'civilization' to the natives together with tuberculosis (Belli, 1897, p. 1296).[7]

Navy Doctors as Promoters of Colonial Expansion

Some of the Navy physicians encouraged colonial expansion and emigration, considered to be a form of colonialism. Filippo Rho enthusiastically claimed that 'After many years of hibernation, Italy—the last in colonial and commercial expansion—is now dispatching her own offspring everywhere in the world, and—for good or ill—setting her foot outside the

[6] Ten years later, Frattini was much more critical of the hygiene conditions in Bombay, including burial practices (Frattini, 1907).

[7] A similar remark about the Salesians was made in the early 1960s by Ettore Biocca, an anti-Fascist parasitologist who had spent the war years in Brazil and organised a scientific expedition to study the Yanomámi in the Amazon: by putting in contact different groups of Amazon natives, they spread tuberculosis, without having the means to fight the disease (Biocca (1965–1966), vol. I, p. 45. See also Cozzoli, 2016).

Mediterranean' (Rho, 1897, p. 406). Navy doctor Mario Mannelli suggested that the Seychelles would be a more suitable anchorage port for Italian ships than Massawa because of the absence of infectious diseases, the presence of a hospital, and their extraordinary nature, which would keep sailors away from alcohol (Mannelli, 1899; on Italian Navy physicians on Massawa sanitary conditions and the measures to be taken, see Bonelli, 2018–9, pp. 101–7). Morocco was also praised by Colorni for its favourable climate and the absence of any endemic infectious diseases, though the physician reported on the miserable health conditions of the indigenous people, the absence of any sanitation measures, and their poor medical knowledge. Colorni criticized the Sultan for his inactivity in improving public health, though he was clearly in favour of European medicine and only trusted European doctors. As noted above, his subjects often went to European doctors but did not follow the prescribed treatments. This clearly highlights the different attitude of the elites and the subalterns towards science and medicine, the same that could be found in southern Italy.

Colorni (1900) emphasized the civilizing mission of the White man, while assessing the colonial value of a region on behalf of the Italian government. Colorni was not an isolated example: the same was done by a score of Navy and Army physicians, including Dr. Salvatore Accurso in Somalia in 1897–1898 (Accurso, 1898), and the physician Eduardo Baccari in the Belgian Congo in 1903, who famously criticized King Leopold's administration and prevented Italian formal expansion in the area (Baccari, 1905; Piccolino, 2007).

Colorni's colonial attitude is made clear by the contrast with the positive reports he made of the British and German colonies in Africa and Asia (Colorni & Rodolfo, 1897). The 'mission' approach was also apparent in Guerra's reports. In 1898, he proposed destroying Mogadishu and rebuilding it according to modern hygienists' criteria (Guerra, 1898), while in later descriptions of the Far East, he praised the colonial work in the British, French, and German concessions and colonies and described Chinese people as dirty and lazy, weakened by the excessive use of opium, implicitly justifying and promoting the penetration of Westerners in China (Guerra, 1900; Giura, 1903a, b). Beyond stressing the contrast between the backward natives and the civilized Europeans, it appears that Italian Navy physicians aimed to learn from other European powers' colonialism. This could be helpful to the Italian colonial administration. Guerra carefully described the city of Saigon, its military hospital—'the most beautiful

I have seen in the Far East' (Guerra, 1900, p. 649)—and the Pasteur Institute. Likewise, Dr. Giura, embarked on the Royal Navy ship *Piemonte*, provided information on the hospitals of the British and German concessions in China, which he considered the 'most complete and beautiful hospitals in the Far East' (see Giura, 1903a, b).

Navy physicians thought of themselves as especially apt for peaceful penetration in the colonized territories, as indigenous people mistrusted missionaries and explorers, but often attended Western doctors (Accurso, 1898). Peaceful penetration was even more important after the Adwa defeat in 1896, with dwindling political support to further colonial military expansion.

Emigration and Colonialism

In 1897, the Royal Navy doctor Filippo Rho wrote one of the first modern comprehensive treatises on tropical medicine: *Predominant Diseases in Warm and Temperate Countries* (see Rho, 1897; on Rho see https://www.treccani.it/enciclopedia/filippo-rho_%28Enciclopedia-Italiana%29/ retrieved on 8 May 2022; Bonelli, 2018–9, pp. 61 and 83–95). Rho was aware of the particular condition of Italy: 'in some provinces, the climate overlaps with the climates of pre-tropical areas in Africa.' Malaria, for example, was not only a colonial disease: it was a national scourge, infesting large coastal areas as well as inland wetlands, with several thousand deaths every year and a severe impact on the national economy.

As a matter of fact, climate was a sort of smokescreen for hiding the social and cultural factors that made those diseases endemic in Italy as well as in the 'colonies.' Rho was hoping for a transformation of tropical countries thanks to the influence of European immigrants, whose role was not only to be the dominant elite, but also the lower classes, including agricultural workers. With a certain pride, he reported the success of immigrants in Brazil: 'after the abolition of slavery, negroes deserted the fields and proved to be helpless as workers in the colonies; Chinese are undesired, Asian Indians performed badly, so that the demand for European workforce is unceasing. European emigration, and especially Italian emigration, is now largely directed to certain intertropical countries in America [...]' (Rho, 1897, p. 730). Italian immigrants were destined to be an alternative to slaves: not exactly a bright future.

Mass emigration, from this perspective, appears to be part of Italian imperial and colonial expansion: the young nation's elite saw mass emigration as a possible solution to the uneven development within the country, mostly in the South—the so-called *questione meridionale.* Rather than obtaining a slow social, economic, and cultural development, subalterns were almost forced to move to a more distant periphery, although they were now charged with imperial ideology and practice.

The masses of people leaving Italy for the Americas were subalterns in their own country. However, while in North America they maintained their condition of subalterns, the matter was rather different in South America. Here they looked at themselves as subalterns with regard to the elites of their motherland, but they were considered colonizers by the local communities. Thus, they were at the same time subalterns and colonizers. The asymmetry was grounded on the difference between the strength of the USA and the weakness of Latin American states (see Garavaglia & Marchena, 2013). While the former easily rejected any attempt by the Italian authorities to interfere in domestic affairs, the latter tolerated the presence of foreign Navy ships in their ports and to some extent pressure from Italian governments.

Navy physicians acted as agents of this complex colonialism, providing hierarchies, collecting information, and establishing practices in order to enforce a colonial relationship with their own compatriots. Doctors on emigration service often called for stricter control over the health status of migrants, who were often rejected upon arrival in North America or repatriated after developing diseases in their new location. An Italian physician practising in the USA—a specialist in tuberculosis—implicitly suggested that poor Italian peasants should not change their lifestyle on arrival in the USA. Several diseases, and especially phthisis, will harm 'the strong countryman more deeply than the pale and haggard city dweller [...] Indian redskins, negroes, several uncivilized populations, and now, the American Italians, are the most irrefutable proof of this' (Stella, 1909, p. 1859). The same constitutional bias is the basis for the poor mental health exhibited by some Italian immigrants after a few months in the USA. A statistical report by the medical colonel Antonio Montano in 1907 highlighted that North American authorities were especially worried about mental health problems that disproportionately (and unsurprisingly, we may add) affected Italian immigrants with respect to 'Nordic races.' Montano's proposal was to widen the medical gaze at departure: immigrants should have had their mental health status checked before boarding, in addition to the screening for infectious diseases. However, according to the Navy medical captain Giuseppe Trimarchi, the psychiatric problem was easily fixable: 'lunatics' will get well soon once they are back in the sunny Italian countryside (Trimarchi, 1909; on Italian colonial psychiatry, see Scarfone, 2014).

In dealing with the 'Other,' the Italian elites apparently adopted the same pattern that John Elliott found for the Britons and the Spaniards who met the American natives in the sixteenth and seventeenth centuries. In his *Empires of the Atlantic,* Elliott argues that the pattern of relationship of the Spanish and the English with the people they met in the New World in the sixteenth and seventeenth centuries was determined by their past experience with the known 'Other,' the Moors and the Irish: the American natives suffered the same segregation policies. Three centuries after, the Italian elites followed a similar path, describing the people they met with the same terms used for the people in the margins of Italy (Forgacs, 2014), the subalterns. Colorni, for instance, defined a nomad population in Somalia as 'gipsies' (Colorni & Rodolfo, 1897, p. 1332). Ten years later, in 1907, Ferdinando Martini, at that time governor of the Italian Colony of Eritrea, compared an Abyssinian tribal leader to a bandit of Sila, a mountain area in the southern Italian region of Calabria (Martini, 1942–3, vol. 4, p. 427). In their African colonies, the Italians ended up adopting the same policies of repression and marginalization they implemented in the motherland.[8]

Medicine participated in this process, though the medical community exploited colonies and emigration in a peculiar way: in declaring malaria, leishmaniasis or cholera 'tropical' or 'colonial' diseases, physicians were 'exoticizing' diseases that were fully present in the homeland and should have been fought with the same means. Physicians and medical researchers, including Royal Navy Physicians, offered help to the Italian colonial enterprise, at the same time carving a space for themselves and for a new discipline—tropical medicine.

[8] In *Prison Notebook 6*, Gramsci had already remarked that the repression of the *Brigantaggio* was carried out along the lines of a colonial expedition (Gramsci, 2007, vol. 3, pp. 3–4).

References

Accurso, S. (1898). Brevi cenni sulle condizioni climatico-igieniche del Benadir e malattie che vi predominano. *Annali di Medicina Navale, 4*, 290–299.

Baccari, E. (1905). Relazione riassuntiva della missione al Congo del Dott. Eduardo Baccari, Capitano Medico nella R. Marina. *Bollettino dell'Emigrazione, 5*, 3–57.

Balduino, C. (1903). La lotta contro la tubercolosi nell'America Latina. *Annali di Medicina Navale, 9*, 838–848.

Ballinger, P. (2020). *The World Refugees Made: Decolonization and the Foundation of Post-war Italy*. Cornell University Press.

Baroni, E. (1899). Analisi chimica di due campioni di veleno per le frecce somale "Ouabaio". *Annali di Medicina Navale, 5*, 364–379.

Baroni, E. (1900). A proposito della Ouabaïna. *Annali di Medicina Navale, 6*, 174–176.

Belli, C. M. (1897). Relazione Sanitaria sulla campagna della Regia Nave 'Colombo' (Anni 1894-5-6). *Annali di Medicina Navale e Coloniale, 3*, 1268–1320.

Biocca, E. (1965–1966). *Viaggi tra gli Indi: Alto Rio Negro, Alto Orinoco. Appunti di un biologo. Consiglio Nazionale delle Ricerche*, 4 volls.

Bodio, L. (1905). Législation et les statistiques comparées de l'émigration et de l'immigration. In *Compte rendu de la session de l'Institut colonial international, tenue à Rome le 25, 26 e 27 avril 1905* (pp. 243–257). Institut colonial international.

Bonelli, C. (2018–9). *Clima, razza, colonizzazione. Nascita e sviluppo della medicina tropicale in Italia* (fine XIX sec.–metà XX sec.). PhD Thesis, University of Rome 'La Sapienza.'

Bonvini, A. (2021). *Risorgimento Atlantico. I patrioti italiani e la lotta internazionale per le libertà*. Laterza

Bottini, C. (1903a). Appunti sull'ospedale civile di Santos (da una relazione del viaggio compiuto dal piroscafo Italia da Genova a Santos e viceversa). *Annali di Medicina Navale e Coloniale, 9*, 333–339.

Bottini, C. (1903b). The Boston City Hospital. *Annali di Medicina Navale e Coloniale, 9*, 270–284.

Bucco, G., & Natoli, A. (1965). *L'organizzazione sanitaria nell'Africa Italiana*. Ministero degli Affari Esteri.

Buonanni, S. (1903). Il nuovo ospedale italiano in Buenos Aires. *Annali di Medicina Navale e Coloniale, 9*, 287–290.

Cappiello, A. (2014). Luigi Bodio: Promoter of the Political and High Scientific Mission of Statistics and Pioneer of the International Statistical Cooperation. *Rivista Italiana di Economia Demografia e Statistica, 68*, 199–205.

Carpi, L. (1874). *Delle colonie e dell'emigrazione d'italiani all'estero sotto l'aspetto dell'industria commercio, agricoltura, e con trattazione d'importanti questioni sociali* (Vol. 4). Tip. Editrice Lombarda.

Colorni, U. (1900). Relazione sanitaria della Missione italiana al Marocco (Ottobre–Dicembre 1899). *Annali di Medicina Navale, 6*, 272–281.
Colorni, U., & Rodolfo, B. R. (1897). Il veleno delle freccie dei Somali o Ouabaïo. *Annali di Medicina Navale, 3*, 1327–1358.
Corbellini, G. (2022). *Storia della malaria in Italia. Scienza, ecologia, società*. Carocci.
Cozzoli, D. (2016). Ethno-biology During the Cold War: Biocca's Expedition to Amazonia. *Centaurus, 58*, 281–309.
Del Boca, A. (1976–1984). *Gli Italiani in Africa Orientale* (Vol. 4). Laterza.
Fantini, B. (1998). Unum facere et alerum non omittere: Antimalarial strategies in Italy. *Parassitologia, 40*, 91–101.
Forgacs, D. (2014). *Italy's Margins: Social Exclusion and National Formation Since 1861*. Cambridge University Press.
Frattini, F. (1907). Le condizioni sanitarie di Bombay ed i pericoli che possono derivare per l'Europa. *Annali di Medicina Navale, 13*, 83–84, 846–52.
Gabriele, M., & Friz, G. (1973). *La flotta come strumento di politica nei primi decenni dello stato unitario italiano*. Ufficio Storico della Marina Militare.
Gabriele, M., & Friz, G. (1982). *La politica navale italiana dal 1885 al 1915*. Ufficio Storico della Marina Militare.
Garavaglia, J. C., & Marchena, J. (2013). *Historia de America Latina de los origenes a la independencia*. II La sociedad colonial ibérica en el siglo XVIII. Crítica.
Gianturco, E. (1899). Sulle odierne cognizioni chimiche intorno al veleno da frecce dei somali "Ouabaio" ed al suo principio attivo "Ouabaina". *Annali di Medicina Navale, 5*, 875–879.
Gianturco, E. (1900). Sull'estrazione della 'Ouabaïna' col metodo Stass-Otto. *Annali di Medicina Navale*, p. 177–8.
Giura, L. (1903a). Appunti sulla colonia inglese di Wei-Hai-Wei. *Annali di Medicina Navale, 9*, 500–502.
Giura, L. (1903b). Appunti sulla colonia tedesca di Kiautschou. *Annali di Medicina Navale, 9*, 503–505.
Gramsci, A. (2007). *Prison Notebooks* (Vol. 3). Columbia University Press.
Gramsci, A. (2021). *Subaltern Social Groups: A Critical Edition of Prison Notebook 25*. Columbia University Press.
Grmek, M. (1973). *Raisonnement experimental et recherches toxicologiques chez Claude Bernard*. Droz.
Guerra, P. (1898). Relazione sulla Campagna della R. Nave "Elba" sulla costa orientale d'Africa. *Annali di Medicina Navale, 4*, 397–415.
Guerra, P. (1900). Relazione climatologica della campagna nell'estremo oriente. *Annali di Medicina Navale, 6*, 642–656.
Hobsbawm, E. J. (1987). *The Age of Empire: 1875–1914*. Weidenfeld and Nicholson.
ICI – Istituto Coloniale Italiano. (1910). *Atti del Primo Congresso degli Italiani all'Estero (Ottobre 1908)*. Coop. Tip. Manuzio.

ICI – Istituto Coloniale Italiano. (1911). *Atti del Secondo Congresso degli Italiani all'Estero (11–20 giugno 1911)*. Tip. Edit. Nazionale.

Labanca, N. (2002). *Oltremare: Storia dell'espansione coloniale italiana*. il Mulino.

Lesch, J. (1981). Conceptual Change in an Empirical Science: The Discovery of the First Alkaloids. *Historical Studies in the Physical Sciences, 11*(2), 305–328.

Leva, F. (1936–60). *Storia delle campagne oceaniche della R. Marina* (Vol. 4). Ufficio Storico della Marina.

Luzzati, A. (1907). La Training School for Nurses nel St. Lukes Hospital. *Annali di Medicina Navale, 13*, 150–155.

Madia, E. (1909–10). Trentotto viaggi in servizio di emigrazione. Studio dell'indole igienico-sanitaria. *Annali di Medicina Navale e Coloniale, 15*, 413–35, 580–603; 16: 290–313.

Maehle, A.-H. (1999). *Drugs on Trial: Experimental Pharmacology and Therapeutic Innovation in the Eighteenth Century*. Rodopi.

Malizia, E. (1903). Note di patologia esotica (da una relazione di viaggio dal piroscafo 'Colombo' da Genova a Parà e Manaos). *Annali di Medicina Navale, 9*, 361–369.

Mannelli, M. (1899). Relazione sulle condizioni climatico-igieniche dei paesi visitati dalla Regia nave "Staffetta" nelle coste orientali d'Africa (1898–1899). *Annali di Medicina Navale, 5*, 487–504.

Manson, P. (1898). *A Manual of Diseases of Warm Climates*. Cassell.

Martini, F. (1942–43). *Il Diario Eritreo* (Vol. IV). Vallecchi.

McIntyre, A. R. (1947). *Curare. Its History, Natural and Clinical Use*. University of Chicago Press.

Molinari, A. (2002). La salute degli emigranti. In Bevilacqua, P., De Clementi, A., & Franzina, E. (Eds.). *Storia dell'emigrazione italiana Arrivi*, vol. II (pp. 377–395). Roma: Donzelli.

Montano, A. (1907). Statistica degli emigrati curati durante l'anno 1906 nelle infermerie di bordo ed appunti sul servizio dell'emigrazione, ricavati dai giornali e dalle relazioni dei medici viaggianti. *Annali di Medicina Navale, 13*, 169–362.

Moroni, E., & Dell'Acqua, F. (1863). *Il Veleno Americano detto curaro. Studi sperimentali, fisiologici e tossicologici applicati alla patologia ed alla terapeutica*. Società per la pubblicazione degli Annali Universali delle Scienze e dell'Industria.

Navarre, P.-J. (1895). *Manuel d'hygiène coloniale: guide de l'Européen dans les pays chauds*. Octave Doin.

Nota, G. (1907). L'università nazionale di La Plata. *Annali di Medicina Navale, 13*, 156–161.

Olivi, G. (1903). L'emigrante italiano a New York (da una relazione di viaggio del piroscafo 'Esperia' in servizio di emigrazione settembre-ottobre). *Annali di Medicina Navale, 9*, 428–440.

Osseo-Asare, A. D. (2014). *Bitter Roots. The Search for Healing Plants in Africa*. The University of Chicago Press.

Packard, R. M. (2007). *The Making of a Tropical Disease: A Short History of Malaria*. The Johns Hopkins University Press

Pasquale, A. (1903). Scopo ed Importanza dell'Igiene navale e coloniale. Discorso Inaugurale della cattedra di Igiene della R. Università di Napoli. *Annali di Medicina Navale, 9*, 371–385.

Piccolino, G. (2007). Il Congo, L'Italia e Leopoldo II del Belgio. Parte II. Nel regno del Caoutchouc e della "Blague" la missione di Eduardo Baccari e le sue conseguenze (1903–1908), in Memorie Scientifiche, Giuridiche, Letterarie, X, Modena, Accademia Nazionale Scienza Letteratura Arti, 2007, pp. 261–318.

Rho, F. (1897). *Malattie predominanti nei paesi caldi e temperati*. Rosenberg & Sellier.

Rho, F. (1903). L'espansione italiana e gli studi di medicina tropicale. *Annali di Medicina Navale, 9*, 387–409.

Ricciardi, T. (2013). *Associazionismo ed emigrazione. Storia delle colonie libere e degli italiani in Svizzera*. Laterza.

Roselli, F. (1907). Il Roosevelt Hospital di New York. *Annali di Medicina Navale, 13*, 714–722.

Ryall, L. (2022). Hidden Spaces of Empire: Italian Colonists in Nineteenth-Century Peru. *Past & Present, 254*, 193–233.

Salvadori, M. (2018). *Storia d'Italia: Il cammino tormentato di una nazione. 1861–2016*. Einaudi.

Scarfone, M. (2014). *La psichiatria coloniale italiana: teorie, pratiche, protagonisti, istituzioni (1906–1952)*. PhD Thesis, Università Ca' Foscari.

Smith, P. (1969). *Arrows of Mercy*. Doubleday.

Snowden, F. (1999). "Fields of Death": Malaria in Italy, 1861–1962. *Modern Italy: Journal of the Association for the Study of Modern Italy, 4*(1), 25–58.

Snowden, F. (2006). *The Conquest of Malaria: Italy, 1900–1962*. Yale University Press.

Stella, A. (1909). La diffusione della tubercolosi fra gli Italiani negli Stati Uniti. *Bollettino dell'Emigrazione, 8*, 1858–1891.

Talamonti, L. (1933). Elmintologia e protozoologia intestinale negli indigeni del basso Webi-Scebeli. *Archivio Italiano Scienze Mediche Coloniali e Parassitologia, 14*, 111–124.

Thomas, B. K. (1964). *Curare: Its History and Usage*. Pitman.

Trimarchi, G. (1909). La pazzia nello stato di New York in rapporto all'immigrazione europea e in ispecie italiana. *Annali di Medicina Navale e Coloniale, 15*, 160–180.

Wagner, F. (2002). *Colonial Internationalism and the Governmentality of Empire, 1893–1982*. Cambridge University Press.

CHAPTER 7

From Universal Rats to Future Jungle Foci: Actors and Places of Plague in Brazil (1899–1940s)

Matheus Alves Duarte da Silva

Like several countries around the world, Brazil faced plague outbreaks throughout the first half of the twentieth century. A more classical historiography examined these epidemics from an institutional angle, showing that, to produce anti-plague vaccines and sera, the Federal Government and the Government of São Paulo created two laboratories in 1899 and 1900, Manguinhos and Butantan, which would ultimately become important centres for microbiology and tropical medicine in Brazil (Stepan, 1976; Benchimol, 1990a; Benchimol & Teixeira, 1993). This panorama was extended in the last decades, and scholars, myself included, started paying attention to other aspects, such as (i) political controversies that emerged following quarantines and other restrictive measures aiming to

The original version of this chapter was previously published without open access. A correction to this chapter is available at https://doi.org/10.1007/978-3-031-38805-7_9

M. A. D. da Silva (✉)
School of Philosophical, Anthropological & Film Studies, University of St Andrews, St Andrews, UK
e-mail: madds1@st-andrews.ac.uk

M. Capocci, D. Cozzoli (eds.), *Empire, Nation-building, and the Age of Tropical Medicine, 1885–1960*, Medicine and Biomedical Sciences in Modern History,
https://doi.org/10.1007/978-3-031-38805-7_7

stop the plague in Brazil, (ii) the circulation of knowledge between the country and other parts of the world also touched by the pandemic, and (iii) the fight against the disease in places such as Rio de Janeiro and São Paulo (Nascimento, 2011; Nascimento & Silva, 2013; Silva, 2018, 2020). However, these new studies privileged one chronological section, between the years 1899 and 1914, and a small part of Brazil, the Southeast. Only recently historians have begun to seriously study the march of plague through the Brazilian hinterland in the 1930s and 1940s, paying attention, for instance, to the institutional history of the Plague National Service (*Serviço Nacional de Peste*), the main actor of a centralized fight against the disease in Brazil. As argued by Simone Luna (2021), this service was created in 1941 not only to fight the plague in rural areas but to avoid the disease becoming endemic in the Amazon and other tropical rainforest zones, which would make its control and eradication virtually impossible.

In this chapter, I propose to bridge these two moments of the history of plague in Brazil and examine how the scourge progressively passed to be framed in the country from a global problem to a potential endemic disease in certain tropical rainforest zones. To describe this transition, I will first focus on the first years of the third plague pandemic (1894–1950), discussing how the rat became a sort of universal villain to be destroyed in ports and maritime structures all around the world. Moreover, I will examine the active role played by Brazil in that global war against domestic rats, by describing deratization measures applied in Rio de Janeiro to fight the plague. In a second moment, I will emphasize some international transformations on plague epidemiology in the 1920s, namely the transition of plague from ports and cities to rural and even wild areas. This internalization of plague gave birth to studies, in several countries, showing that wild rodents could also perpetuate the disease, which coalesced around the idea of sylvatic plague. Created in 1926–1927 by the Portuguese doctor Ricardo Jorge, sylvatic plague was defined as an epizootic among wild rodents living in uninhabited desert-like or steppe-like regions (Jorge, 1927a, 1271; for context, see Silva, 2023). In a third moment, the chapter will focus again on Brazil, reconstructing plague outbreaks on rural villages in the North-East in the 1930s and 1940s and discussing the growing concern that the disease could penetrate wild spaces. This anxiety went in parallel with a reinterpretation of the idea of 'sylvatic plague' in Brazil, where the concept passed to mean jungle plague rather than desert-like plague. I will insist that this U-turn was not a sole semantic transformation, stressing that this new understanding of the 'sylvatic' framed the Amazon and other rainforests as possible new foci of plague in Brazil, albeit not real but forecasted ones.

By following this transition over fifty years, I will argue, firstly, that the plague was mainly seen in Brazil through this whole period as a global disease, mainly because of the widespread presence of domestic rats—vectors, carriers, and later reservoirs of plague. Secondly, I will argue that it was only in the last period examined in the chapter, when sylvatic plague meant 'jungle plague', that it acquired a possible tropical identity. Nevertheless, this tropicalization of plague became a non-realizable prophecy in the long run, because this 'jungle plague' was (and still is) absent from Brazil.

Brazil in a Global War Against the Rat

In 1908, Albert Calmette (1908, 444), director of the Pasteur Institute of Lille, admonished his fellow citizens to 'declare war against the rat'.[1] To Calmette, such enterprise was justified in France and elsewhere because domestic rats[2] were annually responsible for devouring a good amount of food in fields, docks, and ships, being accountable for famine and economic losses. Besides, Calmette (1908, 432) cast rats as responsible for transmitting a plethora of diseases to humans, such as trichinosis, rabies, and especially plague. While the first reason was commonly invoked to justify rat destruction in Europe, being behind rat-catching practices in the nineteenth century and even before (Pemberton, 2014), the fault for spreading diseases was of much more recent date and had its roots in the first years of the third plague pandemic.

It is commonly accepted that the pandemic began in Hong Kong in 1894, an outbreak where the plague bacillus, later christened as *Yersinia pestis*, was described for the first time (Yersin, 1894; Kitasato, 1894). Two years later, the recently identified pathogenic agent was found in India and from there it started spreading over all inhabited continents until the end of the nineteenth century. This first moment of the pandemic was described

[1] All translations from French, Portuguese, and Spanish sources were made by the chapter's author.

[2] Despite being categorized as a sole entity, the use of the term domestic rats in France, Brazil, and elsewhere commonly embraced two different species: the black rat (*Rattus rattus*) and the brown rat (*Rattus norvegicus*).

by Myron Echenberg (2007) as those of the 'plague ports', because of the widespread presence of the disease in those locales, which called for specific sanitary interventions and debates on the international arena.

These initial years of the pandemic were likewise marked by global scientific production, collaboration, and controversies on treatment, vaccination, and transmission of the plague (Silva, 2020). Concerning this third question, domestic rats were seen as possibly implicated in the outbreaks since 1894, partly because plague epizootics among these animals were observed preceding human plague epidemics (Yersin, 1894). Yet their role was a matter of debate. For some doctors, microbiologists, and sanitary authorities, rats died of plague because of soil or environmental contamination, while others pointed to the fact that they were infecting ailments and floors through their dejection, acting then as disease vectors (Silva, 2020, 263–266). In 1898, the French doctor Paul-Louis Simond argued that rat fleas were responsible for transmitting the disease from infected rats to humans (Simond, 1898). Although currently accepted, Simond's hypothesis was met with strong criticism in India and Europe, mainly because of intrinsic problems in his experiment, such as the species of flea used in it and doubts on whether these insects could bite humans (Galli-Valerio, 1900; Indian Plague Commission, 1901, 76–77). The rat-flea theory was only reinforced after the Advisory Committee for the Plague Investigation in India (ACPII), established in 1905, released its first conclusions. Thanks to extensive research conducted in Bombay and Punjab, the ACPII identified the commonest flea of Indian rats—*Pulex cheopis*—and assessed that this insect was responsible for transmitting plague among rats (Advisory Committee for the Plague Investigation in India, 1906; Lamb, 1906, 2). The ACPII also established that this flea could bite humans and live from their blood (Advisory Committee for Plague Investigation in India, 1907, 476). By induction, the committee concluded that the insect was likely responsible for transmitting plague to humans (Lamb, 1906, 8), although this ultimate experiment was not done because of political and ethical concerns (Advisory Committee for Plague Investigation in India, 1907, 472).

Even before these conclusions, sanitary practices targeting domestic rats as implicated in plague transmission mushroomed around the world, from India to the US, passing by Germany and Argentina (Engelmann & Lynteris, 2019; Silva, 2020, 266–269). Brazil was no exception in this fight against the rat. The plague arrived in Brazil at the end of the nineteenth century, touching the ports of Santos, in the State of São Paulo, in

October 1899, and of Rio de Janeiro, in January 1900 (Nascimento & Silva, 2013). Health and hygiene were completely decentralized in Brazil at the time (Hochman, 2013), and partly for this reason, policies diverged in São Paulo and in the federal capital when the plague epidemics started. Informed by research conducted in Asia and by local observation of plague epizootics among rats, São Paulo authorities decided to target these animals along with sick isolation and house disinfection. Therefore, they ordered the poisoning of the sewage system and created a system to reward the hunting and destruction of these animals by the population (Anonymous, 1908). The *Diretoria Geral de Saúde Pública* (DGSP), the federal organism responsible for tackling epidemic diseases in Rio de Janeiro, decided for a more usual approach. Under the command of Nuno de Andrade, the DGSP isolated the sick, vaccinated against plague and disinfected, at times even destroying, houses and goods of those considered infected. Deratization, on the other hand, was seen by Andrade not only as an expensive measure, since the Federal Government would need to purchase rats caught by the population, but also useless, as the role of these animals on the spread of plague was still a matter of controversy (Silva, 2020, 285).

At the end of 1902, this situation began to change after Rodrigues Alves became the new Brazilian president. Willing to transform the capital into a modern display of Brazil, Alves decided for a vast program of urban reconstruction of the old town as well as for a top-down control of the main health scourges affecting Rio de Janeiro: smallpox, yellow fever, and plague (Benchimol, 1990b). To develop this sanitary transformation, Alves chose Oswaldo Cruz as DGSP's new director, a young bacteriologist with training at the Pasteur Institute of Paris. Partly informed by plague control strategies developed in São Paulo and abroad, Cruz created a squadron of rat-catchers, known as *ratoeiros*, whose mission was to purchase rats killed by the population (Cruz, 1904, 19). This was the most infamous measure of Cruz's anti-plague policy, being the target of cartoons suggesting that in place of diminishing the number of rats, it increased it because people were in reality breeding or importing them from other cities (Falcão, 1971, LXV). Rat control in Rio de Janeiro also included the sulfurization of the sewage system by the Clayton machine, house disinfection, and the obligation of owners to refashion their properties, for instance, by filling the gaps between the floors, which was understood as a potential rat's nest (Cruz, 1906; Vasconcellos, 1908, 1910).

This strategy stayed in place until 1912, when plague sensibly decreased in Rio de Janeiro. Because of this continuity, Cruz and his successor at the DGSP, Henrique Figueiredo de Vasconcellos, were successful in producing in scientific reports a strong correlation between continuous destruction of rats and buildings turned rat-proof with a sustained decline in cases and deaths caused by the plague (Vasconcellos, 1908; Anonymous, 1910). This correlation first targeted Brazilian audiences to secure the maintenance of official support to the DGSP. Nevertheless, these data soon circulated to Europe, being used by Vasconcellos to depict the success of the fight against plague in Rio de Janeiro in the well-known journal *Annales de l'Institut Pasteur* (Vasconcellos, 1908). French scholars, such as Simond, likewise mobilized these data from the Brazilian capital for their own projects. After his mission in India, Simond was sent by the French government to study the transmission of yellow fever in Rio de Janeiro, staying there from 1901 to 1905 (Löwy, 1990, 2001). His sojourn coincided with the first years of the new policy implemented by Cruz against the plague, about which Simond would write once back to France (Simond, 1905, 1911, 1913). In those articles and chapters, Simond suggested which measures might be introduced in French colonies, and eventually in France, to control rats and stamp out the plague from places where it was already endemic. Thus, Simond described in detail the fight against the rat in Rio de Janeiro, saying that he observed by himself measures such as sewage sulfurization and the work of *ratoeiros*. More than just communicating it to European audiences, Simond also depicted this policy as a good example of plague control as well as an important proof of the correctness of his hypothesis about plague transmission (Simond, 1911, 472–473 and 514–515).

Rio de Janeiro's rat control records also circulated within the new transnational health organism, the *Office International d'Hygiène Publique* (OIHP), created in 1907. The OIHP, as described by Cueto and colleagues, was both a sanitary consultative agency and an archive, aimed to assemble and share data on, and to propose measures against cholera, yellow fever, and plague (Paillette, 2012; Cueto et al., 2019). Regarding the latter, the OIHP was particularly interested in deratization practices, especially on ships and ports. Coherent with this principle, since 1909 the OIHP assembled the most up-to-date information on such measures (Pottevin, 1910), appealing to Brazil, among other members, to share their experiences (Vasconcellos, 1910). Thereafter, the OIHP *Bulletin* published different tables about the fight against the plague in Rio de

Janeiro, where the correlation between rat control and plague control was reinforced (Anonymous, 1910).

Partly by taking into consideration these data, and partly led by Henry Geddings, US delegate, the OIHP assembly discussed in April 1910 the opportunity of imposing a periodical deratization to all ships allocated to the international trade. This proposal was nonetheless criticized by some European delegates, who stressed the costs of such a measure (Office International d'Hygiène Publique, 1910, 37–43). Therefore, a compromise was reached in the form of an agreement transmitted by the OIHP to its members, which recommended turning ports rat-proof and focusing on rat-catching practices in those locales (Office International d'Hygiène Publique, 1910, 51–52). After WWI, the US brought back the idea of a periodical deratization of all ships. This time the measure gained support in the OIHP and was approved by the International Sanitary Conference of Paris in 1926. From that date onwards, all ships allocated to international trade should pass through a complete deratization every six months, albeit exemptions existed (Office International d'Hygiène Publique, 1928).

In these first decades of the pandemic, a growing consensus was reached that to control local plague outbreaks and limit the international transmission of the disease health authorities should focus mostly on deratization. This global war against the rat seemed possible, and necessary, because of a supposed universality attributed to the animal: not only it existed all over the planet, but it could migrate following maritime traffic. Because of this universality, the fight against the plague in Brazil not only reflected some ideas developed around the world but was also portrayed by local and foreign experts as a kind of proof, and at times a model, that to control the plague one should focus on rat control.

The Invention of Sylvatic Plague

The 1920s represented a turning point in the history of plague globally: ports and coastal cities lost their exclusivity as primary loci of sanitary intervention and international anxiety, and rural and even 'wild' spaces became considered locales where plague could be 'conserved' and eventually spread to humans. These new studies made clear the existence of plague among wild rodents, which coalesced around the concept of sylvatic plague. Despite being more interested in the deratization of ports and ships, the OIHP played a vital role in the emergence of this concept. Since 1910, its delegates were aware that some species of wild rodents, such as the Californian ground squirrel or the Siberian marmot, were susceptible to the plague (Pottevin, 1910, 542; Anonymous, 1912). However, it was only in 1924 that the assembly began to seriously consider this question, starting a global survey on the fauna of rodents and their ectoparasites in relation to the plague (Office International d'Hygiène Publique, 1924, 205). The OIHP invited its members to share their recent or more established research with Ricardo Jorge, the Portuguese delegate. Jorge was chosen to synthesize this material because he proposed the enquiry and was a recognized plague expert (for a contextualization on the role played by Jorge and other actors on the emergence of a modern medicine in Portugal, see Amaral, in this volume).[3] More than thirteen OIHP members—ranging from free countries to colonies—participated in it. Brazil, which was central to previous debates on deratization, did not take part in it (Jorge, 1928).

Jorge's conclusions were presented to his fellows in 1926 and 1927 and published as a book in 1928 (Jorge, 1927a, b, 1928). Given the length of this chapter, I can only focus on Jorge's main argument, which stressed the existence of two types of plague: pandemic and sylvatic (in the original French: *peste pandémique* and *peste selvatique*, respectively) (for a detailed account of the emergence of this concept, see Silva, 2023). Both were caused by *Yersinia pestis* but implicated two different sets of rodents. The pandemic plague was the most widespread, being vehiculated by domestic rats present in almost every port or city on the planet. The sylvatic plague, on the other hand, was 'the work of wild rodents' (Jorge, 1927b, 1271) and, differently from the pandemic plague, had been identified only in four hotspots until 1927: California, the South-African veld, Manchuria, and the Astrakhan region (Jorge, 1927b). To bridge the sylvatic plague foyers together, the Portuguese doctor, who never visited any of those places, insisted on a supposed *emptiness* and *aridness* of the four locales. In his first attempt to define sylvatic plague, Jorge wrote that this 'type of plague exists not in inhabited places but in uninhabited places. Epizootic among wild animals, of desertic habitat, [one] can call it sylvatic plague' (Jorge, 1927b, 1271). Furthermore, in explaining the origins of the four foyers, Jorge again insisted that the sylvatic plague emerged in desertic regions:

[3] Ricardo Jorge (1899) identified the arrival of plague to Porto in 1899, one of the most important outbreaks in Europe during the Third Plague Pandemic.

> The rat plague, penetrating in the hinterland of certain zones of Asia, Africa, Europe, and America, succeeded, by the contamination of wild rodents, very sensitive to the virus, to ignite, in a sustainable way and independent of its first origin, permanent enzootic foci of plague that could be called sylvatic, because of the habitat of its carriers in desertic regions (steppes, Veld, etc.). (Jorge, 1928, 53)

In Jorge's categorization, sylvatic plague was not simply the plague among wild rodents, but the plague of wild rodents living in a specific environment, qualified as 'desertic'. The desert should be understood in a broader sense, meaning both a place without humans (and without domestic rats, by extension) and of scattered vegetation and low rainfall. By arguing for the existence of plague in those locales, Jorge finished showing the existence of global plague reservoirs. Therefore, while Jorge's categorization stressed that a plague control targeting rats in ports and cities should continue and was advisable to tackle the 'pandemic plague,' the complete eradication of the disease seemed vain given the existence of those sylvatic reservoirs.

The sylvatic plague as a new concept slowly gained global traction. In the US, it named a committee created in 1935 led by Karl Meyer that eventually discovered that the Californian hotspot was much larger than originally thought, encompassing large regions of prairies and deserts in different states on the Southwestern United States (Meyer, 1936). In Angola, it was employed by Venáncio da Silva (1935, 1939, 12), director of the *Serviço Permanente de Prevenção e Combate à Peste Bubônica no Sul de Angola*, from 1935 onwards to describe a series of small plague outbreaks that occurred at the border between Angola and South-West Africa (present-day Namibia) caused by wild rodents migrating from the South-African sylvatic plague foyer through the Kalahari Desert. And finally, the British ecologist Charles Elton mobilized the idea of sylvatic plague in 1938 to communicate to British sanitary authorities over the risks of plague seeding in wild animals in Britain via rodents imported by zoos.[4] In parallel to this wide use of the term came some criticism. As remarked by the French plague expert based on Madagascar George Girard, the choice of the word 'sylvatic' was problematic and could lead to mistakes. In fact, the word *selva* in Portuguese, and *selve*, in French, means jungle or forest,

[4] Zoology Elton Archive: Correspondence about sylvatic plague 1938–1939, Bodleian Library, Catalogued Archives and MSS 606612154, CN 222/1/1.

but none of the wild rodents implicated in the sylvatic plague lived in forests or jungle settings, but all inhabited the steppes, remarked Girard. Thus, he preferred the term 'wild rodents plague' (*peste des rongeurs sauvages*) (Girard, 1948), which did not, however, overthrow sylvatic plague in the long run.

These different uses of sylvatic plague, including Girard's criticism, suggest that those authors agreed with Jorge that sylvatic plague touched only wild rodents, being then a kind of peripheral plague, since the commonest plague was transmitted by domestic rats. Jorge's insistence on sylvatic plague being the plague of wild rodents living in desertic habitats seemed less evident, on the other hand. While some authors, such as Girard, understood that this kind of plague was prevalent in a specific type of environment—the steppe—others, like Elton, imagined that sylvatic plague could be brought to Britain, simply meaning thus the plague among wild rodents. Therefore, although they seemed to agree that it was the plague of a very large group of animals, it was still not clear if these rodents should live completely sheltered from the man, especially in arid ecosystems, as Jorge thought, or if, on the contrary, this specificity of the ecosystem was not so central. This ambiguity became particularly evident in Brazil from the 1930s onwards.

Jungle Plague: Sylvatic Plague Transformed in Brazil

While in the 1910s the Brazilian strategy of facing plague in port cities was pictured as a model and proof of modernity, this situation slowly backfired. In the 1920s, scattered plague outbreaks popped up in medium cities and rural villages distant from the coast.[5] This was a problem especially in the North-East of Brazil, a region of the country whose hinterland was faced with recurrent droughts. These plague outbreaks began to be seriously studied from the 1930s onwards, following a broad process of administrative centralization and internalization of public health. This double movement of centralization and internalization started taking roots in 1920 with the creation of the *Departamento Nacional de Saúde Publica* (DNSP), the first sanitary organism with national powers. It was

[5] It does not mean, though, that plague completely disappeared from the coast. For instance, Fortaleza, an important port and capital city of Ceará, was touched in 1934, which created a momentum for those missions. *See* Fialho (1935, 183).

intensified during the Vargas Regime (1930–1945), which witnessed the creation of the first Ministry of Education and Health, in 1930, which absorbed the DNSP and expanded its realm (Hochman, 2013).

A series of decisive missions directed by the DNSP to understand the nature of the plague outbreaks in the North-East was carried out in 1935 and 1936 in the states of Pernambuco, Ceará and Paraíba 'to collect data and study the local conditions aiming to elaborate a general plan to eradicate the plague from the North-East' (Motta, 1936, 210). These missions were conducted by doctors originally based in Rio de Janeiro, who stayed in the North-East for short periods, sometimes when the outbreaks were already receding. Because of that situation, these experts relied not only on their own observations and experiments but on the knowledge shared by local informants, such as farmers and hunters (Motta, 1936, 198). By combining these two sets of knowledge, these plague experts came up, firstly, with a classical conclusion, pointing out that most of the human outbreaks in the North-East were ignited by infected domestic rats. While trying to find food and a nest in poor houses, granaries, and cotton storages abundant in the villages, domestic rats brought plague bacilli with them, which were eventually transmitted to humans by fleas (Parreiras, 1935, 429). Given the role of these well-known vectors, a package of measures already tested in previous outbreaks in Brazil with slight adaptations was proposed. For instance, Marcelo Silva Junior, who studied the plague in Ceará at the end of the year 1935, advised that 'the anti-plague prophylaxis should be oriented in keeping the rat away from the human habitation along with a campaign of deratization exclusively in active plague foci' (Silva Junior, 1936, 183).

Secondly, these experts noticed that some wild rodents, such as *preás* (Brazilian guinea pig, *Cavia aperea*), *punarés* (*Thrichomys apereoides*), and *mocós* (rock cavy, *Kerodon rupestris*), which could be found close to rural villages and were appreciated as bush meat by locals, could be implicated in the plague outbreaks. According to different local actors, human plague outbreaks were preceded by wide annual epizootic waves, which 'destroyed the rodents' and touched even the domestic rat, 'in the farms, villages and, sometimes, even in the cities' (Silva Junior, 1936, 177). Nobody knew with certitude the causes behind these epizootics, but some experts were deeply convinced that it was the plague. This conclusion was, nonetheless, indirect: while those wild rodents were found susceptible to plague in the laboratory, it was not possible to ascertain if they died from the plague in nature (Fialho, 1935, 198). Therefore, their role in the epidemics was deduced because several human cases started after a documented contact with carcasses of rodents and rats killed during these epizootics (Silva Junior, 1936, 177).

A third important point noticed by those experts concerned the fact that before reaching villages and even some cities, these epizootics almost always started in places known as *serras*, where wild rodents were abundant (Fialho, 1935, 197; Silva Junior, 1936, 181). *Serras* could be translated into English as hills; however, this does not do justice to its full meaning in the North-East of Brazil, where *serras* can also mean wild, uninhabited, and even arid places. The aridness was especially stressed by some plague experts. For instance, when wondering which animal could be the plague reservoir in the North-East, Silva Junior argued that it should be a 'vagabond animal or an animal that does not drink water'; therefore, his suspicions lied on the *gato silvestre* (wild cat of unidentified species), a known migrator, or on the *preá*, an animal 'that does not need water to survive' (Silva Junior, 1936, 182). To deal with these animals in the *serras*, some plague experts advised measures different from those centred on rats: instead of a 'derodentization'—an unachievable measure—they suggested that the Federal Government should sponsor surveys to identify the species of rodents of the North-East, study their ectoparasites, and conclude whether these mammals really played a role in human plague outbreaks (Fialho, 1935, 205).

In sum, the studies conducted in 1935–1936 in the North-East of Brazil pointed in two important directions. Firstly, they showed that while rats continued to be likely the main villains of those outbreaks, a pathological role started to be attributed, albeit in an ambiguous way, to a number of wild rodents. Secondly, they pointed to different places where interventions could be directed: both to the *serras*, where the disease seemed to circulate among wild rodents, which pushed for a survey, and to the rural villages, where the plague passed through wild rodents to domestic rats and, eventually to humans, pushing for deratization measures. Because of that new situation, Brazil seemed to mirror other countries, such as South Africa or the US, where plague decreased in coastal cities but survived in rural villages and even in a 'sylvatic' state. However, and quite surprisingly, although many indices seemed to point to the existence of a sylvatic plague foyer as well in Brazil, a different idea of what 'sylvatic' meant emerged in the country and started to inform sanitary strategies against the disease in the North-East from 1941.

This reconstruction of the idea of sylvatic plague in Brazil knew an important moment with the publication of the book *Contribuciones al studio de la peste bubonica en el Nordeste del Brasil*, in 1941, by the Chilean doctor Atilio Macchiavello Varas. Macchiavello had previous experience studying plague and domestic rats in Chile, especially in the port of Antofagasta (Macchiavello Varas, 1932). In July 1939, the Pan-American Health Organization (PAHO), in accord with the DNSP, commissioned him to study the plague in the Brazilian North-East, where he would stay until September 1940, working in the states of Ceará, Paraíba, Pernambuco, Alagoas, and Bahia (Macchiavello Varas, 1941, 9–10). The book, planned as the first part of four, was nothing more than a collection of different reports sent to PAHO and to Brazilian authorities, where Macchiavello and his local collaborators investigated several clinical and epidemiological aspects of the plague.

Macchiavello concluded, as some Brazilian doctors did before him, that domestic rats played the crucial role of spreading and conserving the plague in the North-East, and therefore, the core of any national sanitary strategy should pass by a firm fight against these animals in the form of direct destruction and by turning buildings rat-proof. Nevertheless, plague in Brazil was not exclusively found in these animals, as other scholars had remarked before him. Macchiavello conceded that wild rodents could indeed be infected by the plague and die from it in massive epizootics. However, he found no evidence 'of a primary wild rodents epizootic', but quite the opposite: wild rodents were almost always killed by primary plague epizootics among rats (Macchiavello Varas, 1941, 152–153). Therefore, in Macchiavello's reasoning, the direction of the epizootic wave was different from that described by some Brazilian experts and suggested that the rats were the real plague reservoir in Brazil instead of wild rodents (Macchiavello Varas, 1941, 95). Macchiavello could thus affirm that what existed in Brazil was a kind of rural plague in which wild rodents died almost incidentally. Therefore, a sylvatic infection 'possibly did not exist' in Brazil (Macchiavello Varas, 1941, 101).

Macchiavello knew Ricardo Jorge's works and mentioned them in his 1941 book. Nevertheless, Macchiavello categorized sylvatic in a completely different way from the Portuguese doctor (Macchiavello Varas, 1941, 102). While the latter stressed that deserts and steppes were the natural foci of sylvatic plague, to the Chilean doctor 'sylvatic mean(t) the jungle; a land not cultivated nor dominated by the men, with an exuberant natural vegetation' (Macchiavello Varas, 1941, 16). Therefore, he argued that the foci of a sylvatic plague foyer in Brazil would be the Amazon or other rainforest zones, from where plague was still absent though. Macchiavello did not however vouch that they would stay free of plague forever and urged the Brazilian Government to anticipate this, avoiding this last step of the plague march that, after touching the ports and rural villages, was now in the 'vicinities of the jungle'. Thus, Macchiavello sustained the necessity of a concrete and national system of plague control in the North-East of Brazil, where rats were still the major problem, before plague reached the jungle and its wild rodents, after which its eradication would become a utopia (Macchiavello Varas, 1941, 102–103).

As described in detail by Simone Luna (2021), Macchiavello's reasoning influenced both the creation of the Brazilian Plague National Service in 1941 and its mission. Indeed, the service aimed not only to control rats in cities and villages but also to avoid plague becoming sylvatic in Brazil. This influence became evident, for instance, in a memoir published by Roland Simon, a member of the service, in 1951. According to him, one could devise five phases of the plague march through Brazil: port phase and urban phase were those observed at the beginning of the century and were almost controlled in the country. These were followed by the rural phase and rural-country phase (*fase rural-campestre*, in Portuguese). The rural phase seemed the most present in the North-East of Brazil and was therefore in the centre of the actions of the Plague National Service. The country-rural phase, when both rats and wild rodents were contaminated, was also taking roots in this region, and was then a matter of concern. The last phase, sylvatic plague, defined as the plague among wild rodents living in hyper-humid climate and in places with exuberant and continuous vegetation, did not seem to become a problem in most parts of the North-East given its arid vegetation (Simon, 1951, 25). On the other hand, the Amazon was the 'probable place where the sylvatic plague could appear', wrote Simon (1951, 26). Therefore, he argued that 'it was imperative the intensification of the prophylactic measures in its vicinities to hinder the penetration' of plague in that area (Simon, 1951, 27).

This new understanding of the sylvatic plague did not simply mean a semantic change, but it created a new space of anxiety about the disease, which justified and strengthened the measures applied in the rural villages of the North-East. Different reasons could explain this U-turn in Brazil started by Macchiavello and carried on by actors such as Simon. A plausible one is the parallel emergence of the idea of sylvatic yellow fever in Brazil and other parts of South America (on forests as spaces of epidemiological anxieties but also as locales of scientific studies in Brazil, see Benchimol, in this volume). In 1932, Fred Soper, from the Rockefeller Foundation, and collaborators working in the state of Espírito Santo, Brazil, observed a rural epidemic in which the *Aedes aegypti*, the yellow fever vector in the cities, was absent. Moreover, they noticed that most of the cases occurred in people living in the vicinities of the jungle (Soper et al., 1933). These initial observations were developed in further years, with studies showing that other mosquitoes could spread the yellow fever, and, moreover, that different animals, such as monkeys and small rodents, could be infected and conserve the yellow fever virus. This new research gave birth to the idea of 'jungle yellow fever', in 1935, which pointed to the existence of yellow fever reservoirs among wild animals, especially monkeys, living in the tropical jungle (Soper, 1937, 8). When translated to Portuguese, this jungle yellow fever appeared as *febre amarela silvestre* (sometimes also written as *sylvestre*) (Lane, 1936; Assumpção, 1937). Sylvatic plague was translated in Brazil in very similar terms as *peste silvestre* or *peste selvática* (Silva Junior, 1937; Simon, 1951). Therefore, given the importance of yellow fever in Brazilian tropical medicine, the understanding of the sylvatic focus of yellow fever as the jungle could then have 'cannibalized' the broader understanding of sylvatic plague. Another possible explanation for this new meaning of sylvatic plague in Brazil, which does not exclude the first one, is the etymological connection between the word jungle and sylvatic in Portuguese: *selva* and *silvática or silvestre*, respectively. Therefore, it would be reasonable in Brazil to imagine that a sylvatic disease should be related to the jungle rather than with desertic or steppe-like environments. Anyway, it is worth noticing that this semantic inversion eventually excluded the main areas of the North-East touched by the plague as possible sylvatic plague foci, given their aridness, and explained why the Amazon could be seen as the uttermost sylvatic plague focus in Brazil.

In short, the sylvatic plague that the Plague National Service was trying to avoid since 1941 was no longer the original definition used by Jorge, nor even the idea of plague among wild rodents, but Macchiavello's understanding, which was a completely new definition, that is the plague among wild rodents living in the tropical jungle. Therefore, a new space of plague anxieties emerged in Brazil, which was different from the ports of the beginning of the century, or the rural villages of the 1930s and 1940s, not only by physical and geographic characteristics but also because in that place domestic rats would be absent, and plague would be conserved utterly among wild rodents. Moreover, while in its original definition sylvatic plague was mobilized to describe and act upon present issues in places such as the California foyer, in Brazil sylvatic plague began to mean a kind of prophecy that could be (and had been) avoided: the jungle plague.

Conclusion

This chapter has followed the history of plague in Brazil throughout almost fifty years. It described the disease's path in the country, starting with outbreaks on the coast, followed by its disappearance from that area and emergence in rural villages of the North-East, and finally reconstructing the invention of a feared, but never materialized, last step–plague penetration in the Amazon and other tropical rainforest regions. This internalization of the plague paired the process of centralization of the Federal State and of its public health infrastructure, which progressively expanded from Rio de Janeiro to the Brazilian interior. This expansion was able to identify the disease in new settings, propose sanitary measures against it and ultimately fight it in an increasing national frame. Nonetheless, this national path also had global connections. Indeed, the history of plague outbreaks in Brazil not only mirrored some international experiences, such as the fight against the rat, but created new knowledge through complex dynamics of circulation and translation, being the invention of sylvatic plague as jungle plague, or even Amazonian plague, an important example.

To conclude, we can wonder how plague was understood in Brazil. During the first two decades of the century, it was mostly seen as a global problem, an assumption linked mainly to the widespread presence of domestic rats in every port of the planet. In the 1930s, it came to be understood as a disease of rural spaces, a situation represented in Brazil by the North-East, which found connections with other contexts in tropical and temperate zones of the planet. It was only from the 1940s onwards that the plague started to be assumed as a potential tropical menace when the idea of sylvatic plague was reframed and passed to point to potential tropical rainforest foci. Henceforth, the march of the plague in Brazil was that of a slow 'tropicalization' of the disease in the country.

Funding Research leading to this chapter was funded by the Wellcome Trust [grant ID 217988/Z/19/Z] for the project "The Global War Against the Rat and the Epistemic Emergence of Zoonosis".

References

Advisory Committee for Plague Investigation in India. (1906). I. Experiments upon the Transmission of Plague by Fleas: Part II. Transference from Rat to Rat. *The Journal of Hygiene, 6*(4), 435–449.

Advisory Committee for Plague Investigation in India. (1907). XX. A Note on Man as a Host of the Indian Rat Flea (P. Cheopis). *The Journal of Hygiene, 7*(3), 472–476.

Anonymous. (1908). Resumo dos Trabalhos do Instituto Bacteriológico de São Paulo (1892–1906). Peste Bubônica. *Revista Médica de São Paulo, XI*(4), 77–78.

Anonymous. (1910). Brésil. (Rio-de-Janeiro) Cas et Décès de Peste de 1900 à 1909/Rats Incinérés de 1903 à 1909 et de Maison Rendues Impénétrables aux Rats de 1903 à 1909. *Bulletin Mensuel de l'Office International d'Hygiène Publique, II*(6), 933–934.

Anonymous. (1912). La Peste dans la Série Animale. *Bulletin Mensuel de l'Office International d'Hygiène Publique, IV*(3), 431–480.

Assumpção, L. V. d. (1937). *O Virus Isolado dos Doentes de Febre Amarella Silvestre Durante o Último Surto Epidêmico Verificado no Estado de São Paulo.*

Benchimol, J. (1990a). *Manguinhos do Sonho à Vida: A Ciência na Belle Époque.* Casa de Oswaldo Cruz.

Benchimol, J. (1990b). *Pereira Passos: Um Haussmann Tropical: A Renovação Urbana da Cidade do Rio de Janeiro no Início do Século XX.* Prefeitura da Cidade do Rio de Janeiro, Secretaria Municipal de Cultura, Turismo e Esportes, Departamento Geral de Documentação e Informação Cultural.

Benchimol, J., & Teixeira, L. A. (1993). *Cobras, Lagartos e Outros Bichos: Uma História Comparada dos Institutos Oswaldo Cruz e Butantan.* Editora da UFRJ/Fundação Oswaldo Cruz.

Calmette, A. (1908). Déclarons la Guerre aux Rats. *La Revue du Mois, 3*(28), 432–444.

Cruz, O. (1904). Relatório do Diretor Geral de Saúde Pública. In Brasil. *Ministério da Justiça e Negócios Interiores. Ministro J. J. Seabra. Relatório dos Anos 1903 e 1904 Apresentado ao Presidente da República dos Estados Unidos do Brazil.* Imprensa Nacional.

Cruz, O. (1906). *Peste.* Besnard Frères.

Cueto, M., Brown, T. M., & Fee, E. (2019). *The World Health Organization: A History. Global Health Histories.* Cambridge University Press.

Echenberg, M. J. (2007). *Plague Ports: The Global Urban Impact of Bubonic Plague, 1894–1901.* New York University Press.

Engelmann, L., & Lynteris, C. (2019). *Sulphuric Utopias: A History of Maritime Fumigation. Inside Technology.* The MIT Press.

Falcão, E. d. C. (1971). *Oswaldo Cruz Monumenta Histórica: Tomo 1 a Incompreensão de uma Epoca: Oswaldo Cruz e a Caricatura.* s. n.

Fialho, A. (1935). Estudos sobre a Peste Bubônica no Ceará. *Revista de Hygiene e Saúde Pública, IX*(6), 180–205.

Galli-Valerio, B. (1900). Les Puces des Rats et des Souris Jouent-Elles un Rôle Important dans la Transmission de la Peste Bubonique? *Centralblatt fur Bakteriologie, Parasitenkunde und Infektionskrankheiten, XXVII*(1), 1–4.

Girard, G. (1948). Sur un Point de Terminologie: L'expression Peste Selvatique ou Sylvatique est Fondamentalement Erronée. *Bulletin de la Société de Pathologie Exotique, 41*, 15–16.
Hochman, G. (2013). *A Era do Saneamento: As bases da política de Saúde Pública no Brasil* (3rd ed.). Hucitec.
Indian Plague Commission. (1901). *Report of the Indian Plague Commission with Appendices and Summary* (Vol. V). Eyre and Spottiswoode.
Jorge, R. (1899). *A Peste Bubônica no Porto-1899.* In *Seu Descobrimento. Primeiros Trabalhos.* Repartição de Saúde e Higiene da Câmara do Porto.
Jorge, R. (1927a). Les Faunes des Rongeurs et des Puces dans leurs Rapports avec la Peste (Première Partie). *Bulletin Mensuel de l'Office International d'Hygiène Publique, XIX*(8), 1094–1109.
Jorge, R. (1927b). Les Faunes des Rongeurs et des Puces dans leurs Rapports avec la Peste (Deuxième Partie). *Bulletin Mensuel de l'Office International d'Hygiène Publique XIX, 9*, 1257–1288.
Jorge, R. (1928). *Les Faunes Régionales des Rongeurs et des Puces dans leurs Rapports avec la Peste: Résultats de l'enquête du Comité Permanent de l'Office International d'hygiène Publique 1924–1927.* Masson et cia.
Kitasato, S. (1894). The Bacillus of Bubonic Plague. *The Lancet, 144*(3704), 428–430.
Lamb, G. (1906). The Advisory Committee for the Plague Investigation in India: Third Progress Report to the Committee. SA/LIS/J.3: Box 6. Wellcome Library.
Lane, J. (1936). Notas Sobre Investigações Entomológicas em Localidades Onde Houve Febre Amarella Sylvestre em São Paulo. *Archivos de Hygiene e Saude Publica, I*(2), 125–133.
Löwy, I. (1990). Yellow Fever in Rio de Janeiro and the Pasteur Institute Mission (1901–1905): The Transfer of Science to the Periphery. *Medical History, 34*(2), 144–163. https://doi.org/10.1017/S0025727300050626
Löwy, I. (2001). *Virus, moustiques et modernité: la fièvre jaune au Brésil, entre science et politique.* Éd. des Archives Contemporaines.
Luna, S. (2021). *O Serviço Nacional de Peste e o Controle Da Peste Bubônica No Nordeste Brasileiro (1941–1956).* Master's Thesis, Fundação Oswaldo Cruz/ Casa de Oswaldo Cruz.
Macchiavello Varas, A. (1932). Estudio Sobre las Ratas de Puerto de Antofagasta en Relacion con las Epidemias de Peste Bubonica. *Revista del Institut Bacteriologico de Chile y de la Sociedad Chilena de Microbiologia e Higiene, III*(1), 39–72.
Macchiavello Varas, A. (1941). *Contribuciones al Estudio de la Peste Bubonica en el Nordeste del Brasil.* Pan American Sanitary Bureau.
Meyer, K. F. (1936). The Sylvatic Plague Committee. *American Journal of Public Health and the Nation's Health, 26*(10), 961–969.
Motta, M. d. C. (1936). O Problema da Peste no Estado da Parahyba. *Archivos de Hygiene, VI*(1), 187–210.
Nascimento, D. R. d. (2011). La Llegada de la Peste al Estado de Sao Paulo en 1899. *Dynamis: Acta Hispanica Ad Medicinae Scientiarumque Historiam Illustrandam, 31*(1), 65–83. https://doi.org/10.4321/S0211-95362011000100004

Nascimento, D. R. d., & Silva, M. A. D. d. (2013). 'Não é Meu Intuito Estabelecer Polêmica': A Chegada da Peste ao Brasil, Análise de uma Controvérsia, 1899. *História, Ciências, Saúde-Manguinhos, 20*(suppl 1), 1271–1285. https://doi.org/10.1590/S0104-59702013000400010

Office International d'Hygiène Publique. (1910). *Session Extraordinaire d'avril 1910 du Comité Permanent de l'Office International d'Hygiène Publique: Procès-Verbaux Des Séances.* Imprimerie-Librarie Reunies Martinet Dr.

Office International d'Hygiène Publique. (1924). *Session Extraordinaire d'avril-mai 1924 du Comité Permanent de l'Office International d'Hygiène Publique: Procès-Verbaux Des Séances.* Imprimerie Nationale.

Office International d'Hygiène Publique. (1928). Modèle de Certificat de Dératisation (Ou d'exemption de Dératisation) Adopté par l'Office International d'Hygiène Publique dans sa Session de Novembre 1927. *Bulletin Mensuel de l'Office International d'Hygiène Publique, XX*(2), 295.

Paillette, C. (2012). Épidémies, Santé et Ordre Mondial. Le Rôle des Organisations Sanitaires Internationales, 1903–1923. *Monde(s), 2*(2), 235–256. https://doi.org/10.3917/mond.122.0235

Parreiras, D. (1935). Notas e Estudos Sobre a Peste no Nordeste do Brasil. *Problema Nacional*, 429–433.

Pemberton, N. (2014). The Rat-Catcher's Prank: Interspecies Cunningness and Scavenging in Henry Mayhew's London. *Journal of Victorian Culture, 19*(4), 520–535. https://doi.org/10.1080/13555502.2014.967548

Pottevin, H. (1910). La Dératisation: Rapport sur l'État Actuel des Méthodes Applicables à la Destruction des Rongeurs et de leurs Parasites Établi pour Être Presenté au Comité Permanent de l'Office International d'Hygiène Publique par M. Le Docteur Henri Pottevin, Secrétaire Général de l'Offfice. *Bulletin Mensuel de l'Office International d'Hygiène Publique, II*(4), 542–612.

Silva, F. V. d. (1935). Relatório sobre a Peste Selvática em Angola. Elaborado a Pedido do Prof. Dr. Ricardo Jorge'. Espólio 18, caixa 18, maço 5. Biblioteca Nacional de Portugal – Reservados.

Silva, F. V. d. (1939). *Animais Bravos nas suas Relações com os Problemas Sanitários do Sul de Angola*. Imprensa Nacional.

Silva, M. A. D. d. (2018). From Bombay to Rio de Janeiro: The Circulation of Knowledge and the Establishment of the Manguinhos Laboratory, 1894–1902. *História, Ciências, Saúde-Manguinhos, 26*(2), 1–19. https://doi.org/10.1590/s0104-59702019000200001

Silva, M. A. D. d. (2020). *Quand la Peste Connectait le Monde: Production et Circulation de Savoirs Microbiologiques entre Brésil, Inde et France (1894–1922).* Thèse de Doctorat, Ecole des Hautes Etudes en Sciences Sociales.

Silva, M. A. D. d. (2023). Between Deserts and Jungles: The Emergence and Circulation of Sylvatic Plague (1920–1950). *Medical Anthropology*. https://doi.org/10.1080/01459740.2023.2189110

Silva Junior, M. (1936). Peste no Ceará. *Archivos de Hygiene, VI*(1), 155–186.

Silva Junior, M. (1937). Peste Silvestre no Ceará. *Archivos de Hygiene, VII*(1), 95–97.

Simon, R. (1951). *Verificação da Sensibilidade dos Roedores da Região Neotrópica.* Gráfica Debret.
Simond, P.-L. (1898). La Propagation de la Peste. *Annales de l'Institut Pasteur, XII*(10), 625–687.
Simond, P.-L. (1905). *La Question du Véhicule de la Peste (Extrait de La Revue Médico-Chirurgicale du Brésil et de l'Amérique Latine, N° 5 de 1905).* Besnard Frères.
Simond, P.-L. (1911). Peste. In A. Chantemesse & E. Mosny (Eds.), *Traité d'hygiène* (Vol. 17, pp. 429–535). J.B. Baillière et Fils.
Simond, P.-L. (1913). Peste. In C. Grall & A. Clarac (Eds.), *Traité de Pathologie Exotique, Clinique et Thérapeutique. Maladies Parasitaires, Peste* (pp. 453–648). J.B. Baillière et Fils.
Soper, F. L. (1937). The Newer Epidemiology of Yellow Fever. *American Journal of Public Health and the Nations Health, 27*(1), 1–14. https://doi.org/10.2105/AJPH.27.1.1
Soper, F. L., Penna, H., Cardoso, E., Serafim, J., Frobisher, M., & Pinheiro, J. (1933). Yellow Fever Without Aëdes Aegypti. Study of a Rural Epidemic in the Valle do Chanaan, Espirito Santo, Brazil, 1932. *American Journal of Epidemiology, 18*(3), 555–587. https://doi.org/10.1093/oxfordjournals.aje.a117967
Stepan, N. (1976). *Beginnings of Brazilian Science: Oswaldo Cruz, Medical Research and Policy, 1890–1920.* Science History Publications.
Vasconcellos, H. F. d. (1908). Prophylaxie de la Peste à Rio de Janeiro. *Annales de l'Institut Pasteur, XXII*(10), 819–831.
Vasconcellos, H. F. d. (1910). La Prophylaxie de la Peste à Rio de Janeiro. In *Session Extraordinaire d'avril 1910 du Comité Permanent de l'Office International d'Hygiène Publique: Procès-Verbaux des Séances* (pp. 11–31). Librairies-Imprimeries Réunis Martinet Dr.
Yersin, A. (1894). La Peste Bubonique à Hong-Kong. *Annales de l'Institut Pasteur, VIII*(9), 662–667.

CHAPTER 8

Anti-Fascist Medicine and the International Peace Campaign Against Urban Raids in Spain and China, 1936–1939

Carles Brasó Broggi

Introduction

In the spring of 1938, the cities of Madrid, Barcelona, Wuhan, and Guangzhou were suffering from the same disease. It was a new kind of violence that stroke these urban citizens, creating a state of sanitary emergency. The sound of air raid alarms alerted the population that had to run and seek shelter as enemy planes approached. Next, bombs fell randomly destroying houses and killing or wounding their inhabitants without

This research is supported by a Ramón y Cajal fellowship (RYC 2018-024078-I), funded by the Spanish Ministerio de Ciencia e Innovación and the European Social Fund. The author is also a researcher of ALTER group at the Universitat Oberta de Catalunya (project TACK: PID2019-108637GB-I00).

C. Brasó Broggi (✉)
Faculty of Arts and Humanities, Open University of Catalonia, Barcelona, Spain

M. Capocci, D. Cozzoli (eds.), *Empire, Nation-building, and the Age of Tropical Medicine, 1885–1960*, Medicine and Biomedical Sciences in Modern History,
https://doi.org/10.1007/978-3-031-38805-7_8

distinctions. Local hospitals were suddenly flooded by people and, sometimes, were bombed as well. During three days between March 16 and 18 in 1938, air raids killed around a thousand citizens from Barcelona, while more than 2000 were wounded (Solé i Sabaté & Villarroya, 2003; Gallego Vila, 2017; Arañó & Capdevila, 2019). In June, the mayor of Guangzhou, Zeng Yangfu reported the same approximate number of casualties as well as more than 5000 wounded after twenty consecutive air raids. The Chinese mayor urged to organize an international conference to discuss this new form of violence.[1]

Air raids appeared in the first decades of the twentieth century in both the center of Europe and the peripheries of the imperial dominions, but their impact was relatively limited until the mid-1930s (Holman, 2014; Grayzel, 2012; Omissi, 1990). After the traumatic events of the First World War, the League of Nations tried to enforce the Hague conventions (1899 and 1907) that banned military attacks against open cities and, specifically, of throwing objects from the air pointing at non-military targets (Grayzel, 2012).[2] However, this practice was soon used again by the European air forces to secure colonial control over entire territories and populations ranging from Morocco, Sudan, Arabia, and India (Omissi, 1990).

In 1938, the call of the mayor of Guangzhou was acknowledged by the Rassemblement Universel pour la Paix/International Peace Campaign (hereafter RUP/IPC), an international pacifist organization founded by French and British political activists to claim the inviolability of the international rules set by the Hague conventions and further dispositions of the League of Nations. In 1935, the RUP/IPC initiated its activities as a reaction to the countries that were not abiding this international framework, mainly Japan, Germany, and Italy, and focused on criticizing the Italian air raids against open cities of Abyssinia that were clearly attacking civilian targets.[3] It did not challenge, however, air raids in French and

[1] "Les bombardements des villes ouvertes; L'appel du maire du Canton", Geneva, June 8, 1938, International Institute of Social History (thereafter IISH), Amsterdam, Rassemblement Universel pour la Paix Archives (RUPA), hereafter IISH/RUPA, No. 176, p. 8.

[2] Rassemblement Universel pour la Paix, preparation for September congress, May 1, 1936; and Jean Plot, "Il faut que réussisse le Rassemblement Universel pour la Paix", January 1, 1936, press article at Œuvre, Archives Nationales, Surété Nationale, 20010216/183 Dossier 6573 Comité Français du RUP, 1936–37.

[3] René Maublanc, "Une victoire sur l'esprit de la guerre". In *Les peuples respondent Oui! Le Monde entier pour la paix*. Paris, 1936: Rassemblement universel pour la paix. Publication available at (January 25, 2022): https://sourcespaix.hypotheses.org/bdic_fdr0716_07_38_001

British colonial dominions, at least in the beginnings of its activities. However, in 1937, as the threat of a global war became more prevalent, the RUP/IPC called for the abolition of the aerial warfare and the creation of an international air police force.[4]

This chapter aims to study this social movement as a global phenomenon—that embedded some of the features of the late imperialist era that started in the last decade of the nineteenth century—and its impact upon international medicine. Even though it had clear similarities with the Red Cross movement (Hutchinson, 1996), the RUP/IPC aimed to distinguish and criticized the Swiss health organization for being too lenient with imperialistic interests. This was also the age of tropical medicine and was characterized by the enlargement of imperial dominions through the intensive use of technological and scientific innovations: from quinine and the yellow fever vaccine to machine guns and airplanes (Omissi, 1990; Neill, 2012). Like tropical medicine, air raids symbolize the entanglements between science and colonial violence: while air raids were used as a technological device to patrol border areas, extend imperial dominions, and punish the civilian population, this practice brought some unintended consequences, such as the emergence of an international movement of resistance as well as new streams of scientific research focusing on treating the victims.

Blood banks, ambulances, and the know-how of air-raid shelters were improved in the Spanish Civil War, when air raids became widely and massively used for the first time, mainly by Italian and German air forces who supported the Spanish rebel generals.

One of the major innovations that took place in Republican Spain was the creation of blood transfusion centers that could extract, preserve, transport, and deliver blood from a list of donors (volunteers) to the wounded in the front or in the city hospitals (Palfreeman, 2015). Thus, blood transfusions could take place immediately on the spot, without having to find a donor with a same blood type. This was due to the leadership of two doctors, Frederic Duran Jordà in Barcelona and the Canadian doctor Norman Bethune in Madrid when both cities were suffering from air raids. The Canadian was a well-known thoracic surgeon that arrived in

[4]Correspondance with the International Peace Campaign, League of Nations Archive, R4226/7B/29342/664, May–June 1937 (available online, last access July 11, 2022): https://archives.ungeneva.org/correspondance-avec-le-rassemblement-universel-pour-la-paix

Madrid in November 1936, when the city was besieged and frequently bombed. In less than six months of hectic activity, Bethune created the Canadian Blood Transfusion Service (Servicio Canadiense de Transfusión de Sangre), a pioneering center of blood extractions, preservation, and transportation. He was also involved in broadcasting the consequences of aerial bombardments: first in a documentary about air raids in Madrid (entitled "Madrid bajo las bombas") and second in an illustrated book about the bombings of Malaga (Bethune, 1937).[5] Air raids in Spain were widely reported as a new form of total war.

Ambulances and sanitary cars were also improved. Duran Jordà, besides formulating a method to preserve blood with sodium citrate (that allowed indirect transfusions), adapted a refrigerated car to transport preserved blood, so his lab in Barcelona could provide blood to both the city hospitals that received air-raid victims and the wounded soldiers at the frontline (Broggi, 1997; Grífols, 1997). The ambulances equipped with surgical equipment were ameliorated and used massively after the first experiments in the French army during the First World War. The air raids against cities and hospitals brought new developments in the manufacture of automobiles equipped with surgical capacity, the so-called "auto-chirs" (auto-chirurgicales) (Olier, 2002; Brasó Broggi, 2022).

Furthermore, air raids against urban civilians transformed the management of hospital emergencies. According to the experience of Josep Trueta, the head surgeon of Barcelona's General Hospital, it was essential to have good rescue squads that could take the wounded to the hospital as fast as possible, without further classification process. Rescue squads collaborated with the ambulance services and the network of the city's general hospitals, considering the number of available beds and the operating list of the teams of surgeons that worked in shifts. Wounds were classified according to the type of bombs that were used and the affected parts of the body: multiple injuries (31.6 per cent), lower limbs (22.3 per cent), arms and shoulders (14.7 per cent), head and face (13.2 per cent), abdomen (9.1 per cent), and thorax and chest (8.1 per cent).[6] When Trueta migrated to London in February 1939, his methods became an international standard, along with a similar technique developed in Madrid by

[5] The first one is lost. See letter of Joaquin d'Harcourt to R. J. Stewart, in The Taminent Library & Robert F. Wagner Labor Archives, Archives of the Abraham Lincoln Brigade Archives (ALBA), Fredericka Martin Archives, folder Bethune.

[6] Josep Trueta Raspall, "Opinons formed from actual experience of Aerial bombardment of Barcelona", The National Archives (Kew Gardens, London), FD 1/5372, pp. 35–55.

Manuel Bastos Ansart, which would be known as the "Trueta method" (Trueta, 1939; Rodrigo, 2009), the "Bastos-Trueta method" or simply the "Spanish method" of healing war wounds (Moral Torres, 1986).

By the end of the 1930s, most of air raids had no clear military purpose but to break the morale of the civilian population (the so-called moral effect), but as it had happened during the First World War, the victims of air raids usually transitioned from a process of terror and panic, at first, to adaptation and, later on, to active resistance (Omissi, 1990). The campaigns of the RUP/IPC show the interplay between colonial, national, and international public health (as it is underlined in the introduction by Capocci and Cozzoli) and how resistance to air raids was organized globally into a movement that transitioned from international pacifism to anti-fascism and anti-imperialism. Thus, air raids became associated with imperialist and fascist violence (Amrith, 2006; Louro, 2018).

The RUP/IPC Facing Fascism and Imperialism

The RUP/IPC was an international social movement that gathered individuals and institutions from different political backgrounds and ideologies: the presidents of the organization were Lord Cecil, a conservative British politician, and one of the architects of the League of Nations, and the leftist minister of the French Popular Front, Pierre Cot. It also integrated statesmen from other countries, like Manuel Azaña and Edvard Beneš (presidents of Spain and Czechoslovakia); anti-imperialist, pacifist and nationalist leaders, like Indians Jawaharlal Nehru and Mahatma Gandhi; Communist Party members, like Marcel Cachin (founder of the French Communist Party); Comintern agents (Louis Dolivet); social democrats and socialists (Louis de Brouckère, Solomon Grumbach, Julio Álvarez del Vayo); religious personalities (Jules Jézéquiel); army generals (Paul Pouderoux); sportsman (Philip Noel-Baker); feminist activists (Gabrielle Duchêne); and also intellectuals, like the Chinese education reformer Tao Xingzhi, the Nobel prize winner Rabindranath Tagore, or the well-known poets Paul Valéry and Federico García Lorca (Mazuy, 1993; Ren, 2018).[7] It was a progressive catch-all movement that became identified with global anti-fascism.

[7] A/S d'une nouvelle Enterprise pacifiste et défaitiste, Activité du RUP, Paris, 27 February, 1936, Archives Nationales, Surété Nationale, 20010216/183 Dossier 6575; See also *Les peuples respondent Oui! Le Monde entier pour la paix.* Paris, 1936, Rassemblement universel pour la paix. Publication available at (January 25, 2022): https://sourcespaix.hypotheses.org/bdic_fdr0716_07_38_001

The RUP/IPC was organized in three centers, Paris, London, and Geneva, and its first international event was celebrated in Brussels in September 1936: the Universal Congress for Peace. The congress had a profound international impact as it coincided with a massive mobilization of international support toward the Spanish Republican government after the coup of the rebel general Francisco Franco. The rapid advance of Franco that summer, helped by the German and Italian governments of Hitler and Mussolini, and the massive bombings of Madrid in November symbolized how pacifism had to actively stand against the aggressiveness of some countries, while, at the same time, criticizing the non-belligerence and non-intervention of the liberal governments of France, United States, and Great Britain.[8] Indeed, the leading members of RUP/IPC (Pierre Cot, Lord Cecil) came from the diplomatic elite of the League of Nations that had been disillusioned by its impotence in sanctioning the aggressions and the lack of enforcement of the rules of collective security (Farrugia, 2011).[9]

The conference of Brussels gathered more than 5000 attendants from 35 countries. It collected endorsements from other organizations that, according to the organization, included a total of four million members and affiliates (Mazuy, 1993). The conference was followed by massive demonstrations and debates about the menace of war and how to "fight for peace."[10] There were commissions on medicine, psychology, and scientific cooperation and a specific commission on aviation, where the new forms of aerial warfare were discussed.[11] Further meetings of the medical commission would integrate experts in aeronautics, as well as doctors, to discuss about air raids and their consequences.[12]

[8] *Les peuples respondent Oui!* Op. cit., p. 8.

[9] Pierre Cot, speech in the Conférence universelle d'action pour la paix et contre les bombardements des villes ouvertes. In *Au secours des victimes de l'agression, débats de la Conférence universelle d'action pour la paix et contre les bombardements des villes ouvertes.* Paris, les 23 et 24 juillet 1938. Geneva, 1938: le Rassemblement universel pour la paix. Publication available at Bibliothèque Nationale de France.

[10] Friedrich Stampfer, "Il faut dire aux peuples la vérité", Archive Nationales, Surété Nationale, 20010216/183 Dossier 6575; *Les peuples respondent Oui!* Op. Cit., p. 15.

[11] "Resolutions", Medical and Psychological commission, Scientific Commission and Aviation Commission, Archives of the League of Nations, R5160/13/23643/678/Jacket2, International Peace Campaign, pp. 134–39.

[12] Les comissions s'organisent, *Rassemblement,* December 1936, No. 3, p. 4; League of Nations Archive, R5564/19/26432, available online (July 11, 2022): https://archives.ungeneva.org/q2tg-xwcf-tst4

The organization defined itself as pacifist and defeatist, as a global war was perceived to be imminent and unavoidable. According to French general Paul Pouderoux, who wrote a pamphlet that became a kind of a founding statement of RUP/IPC, both the defensive and aggressive tactics against air raids, while unavoidable, were terrible and useless.[13] This vision coincided with some apocalyptic theories of the interwar period that perceived air raids as a fatal weapon that would decide the fate of the future wars in only several days (Holman, 2014). In the spring of 1939, in the conference at the Royal Society of Medicine, Doctor Trueta had to refute the British previsions of hundreds of thousands of victims in air raids (if so, wars would only be a matter of days): victims of air raids in London could be estimated in around ten thousand at most; so, air raids were an important new aspect of war, but they would not be as decisive as it was thought in the beginning (Rodrigo, 2009).

Meanwhile, the RUP/IPC tried to merge the international liberal pacifism of Geneva, the anti-fascist struggle that symbolized the French and Spanish popular fronts, and the anti-imperialist movements that emerged in the 1930s, especially in India and China. It was a complex mix of ideas that ultimately doomed the project to failure. This debate took place in the intellectual centers like Sorbonne University, but also in small towns and third tier cities. In October 1936, in the town of Fougères, 450 attendants from labor unions and leftist political parties acclaimed the Abyssinian and Spanish representatives introduced by the secretary of the Communist youth league.[14] Meanwhile, the mayor, the local church, and the labour unions of Bolton, an industrial district near Manchester, organized a solidarity week for peace and in support of both countries. These campaigns were both local and global and attracted the attention of other activists in the United States, Mexico, the Soviet Union, Poland, Czechoslovakia, Rumania, India, Egypt, and China. While adherents in Western countries came from both the diplomatic circles and the left, the range of sympathies in colonized, semi-colonized, or war-threatened territories like Spain, Czechoslovakia, India, or China was broader, involving nationalists and liberals. In China, for instance, the nationalist general

[13] Libelle de Pouderoux, in A/S d'une nouvelle Enterprise pacifiste et défaitiste, Activité du RUP, Paris, 26 November 1935, Archive Nationales, Suréte Nationale, 20010216/183 Dossier 6575.

[14] Le commissaire de police de la ville de Fougeres à monsieur sous-prefet de Fougeres, 10 October 1936, Archive Nationales, Surété Nationale, 20010216/183 Dossier 6573, p. 66–67.

Feng Yuxiang and the minister of finance of the Kuomintang, H. H. Kung (Kong Xiangxi), promoted the activities of the RUP/IPC.[15]

The RUP/IPC also received the active support of the Communist international, the Comintern (Schauff, 2006; Buchanan, 2012). Indeed, it was influenced by international communism, especially from the sector of Willi Münzenberg (Mazuy, 1993; Louro, 2018). Willi Münzenberg had organized an international conference against imperialism (the Anti-imperialist league) in 1927 and a successful aid campaign to China and Syria in 1925 (Petersson, 2013). When he moved to Paris after the rise of Hitler in Germany, he continued his political activism, especially focusing on the war in Spain, the recruitment of the International Brigades, and the medical aid campaigns in support of the Spanish Republic (Brasó Broggi, 2022). In the autumn of 1937, the Comintern launched a campaign to support Spain and China, perceiving that both were defensive wars. According to this theory, defending the governments of Spain and China was defending the cause of world peace against the international fascist threat.[16] The RUP/IPC was heavily influenced by this vision of the Comintern as well as by its use of international propaganda.

However, the RUP/IPC became too massive, too global, and too heterogeneous to be a mere instrument of the Soviet Union or the Comintern (Mazuy, 1993). The two-year period of 1936 and 1938, coinciding with the rise of popular fronts in France, Spain, and China, was the heyday of the RUP/IPC, a "golden age" that was short-lived (Farré, 2007). The Conference on air raids in open cities that took place in Paris in July 1938, as a response to the call of the mayor of Guangzhou, was a culminating point. It was a diverse and international event that gathered more than 1000 attendants, representing 34 countries and 24 international organizations. While Jacques Duclos spoke in the name of the French Communist Party, other leaders that were clearly not communist sympathizers, like Chinese general Chiang Kai-shek or US Republican politician John Foster Dulles, also sent messages of support.[17] Therefore, the RUP/IPC had a

[15] "Une campagne globale pour la défense de la paix", dossier, *Le Peuple*, 16 October 1936, Archive Nationales, Surété Nationale, 20010216/183 Dossier 6573, p. 64.

[16] "On aid for the Chinese and Spanish people", October 1, 1937, Russian State Archive of Socio-Political History (RGASPI), Archives of the Executive Committee of the Comintern, 495/18/1225. Archive available online (July, 11, 2022): http://sovdoc.rusarchives.ru/sections/organizations//cards/241917

[17] *Au secours des victimes de l'agression,* op. cit.

communist influence but went beyond the definition of a communist movement.

In that conference, the Indian nationalist leader Nehru compared the air raids in Spain and China—Nehru had experienced them in Barcelona and Chongqing—and the British punishing campaigns that, since the end of the First World War, had ravaged not only the North Western regions of India, but also Iraq, Aden, Egypt, Sudan, Syria, and Somaliland. According to Nehru, "We can see today the fascist aggression pushing the world to war and, justifiably, we condemn it and we seek to fight against it. However, even though fascism is a recent phenomenon in the West, we have experienced it since long ago, but in a different form and another name: imperialism."[18] Thus, air raids were a kind colonial disease that, in contrast to other tropical maladies, was artificial and consciously provoked by aggressive nationalism and its intensive use of technology. According to Nehru, there was a clear continuity between European imperialism in the periphery and the rise of fascism in Europe: they were actually the same and had to be countered by an international civilian resistance movement (Louro, 2018). This statement also unveiled the double standard of the organization, its French and British origins, and the long-established cultural representations of the colonial world that justified air raids in Iraq but not in Europe (Satia, 2006).

Furthermore, the RUP/IPC and the broad political alliances of the popular fronts also decayed for other reasons. In September 1939, the clash of the Second World War in Europe ended with the discussions about active pacifism, a concept that suddenly appeared naïve and useless. Furthermore, the Stalinist purges and the pacts of Munich and the Ribbentrop-Molotov endangered the coexistence of communist, leftist, nationalists, and liberal affiliates. The fact that the RUP/IPC criticized the Ribbentrop-Molotov pact is indicative of its independency regarding the political mandates of Moscow. The situation also spurred fears of a massive bombing in the European capitals like London or Paris, whose population could be massively murdered following the catastrophic theory of air raids (the so-called "knockout blow theory"), a theory that would not be materialized until the end of the conflict in August 1945 (Holman, 2014). The events of 1939 broke the inner coherence of the organization that languished until practically disappearing while raids against cities became

[18] Transcription of the speech of Nehru, in *Au secours des victimes de l'agression*, op. cit., p. 67.

widespread (Mazuy, 1993). Due to the short duration of RUP/IPC and its failed mission, the academic literature dealing with this organization is scant (Mazuy, 1993; Farrugia, 2011; Ren, 2018).

The Spanish and Chinese Campaigns

While other parts of the RUP/IPC deserve further investigations, this chapter focuses on the peak years of RUP/IPC, between 1937 and 1938, when both the Spanish Civil War and China's war with Japan coexisted. After the Marco Polo Bridge incident of July 1937, Japan started a full-fledged invasion on China and air raids became a usual practice of the Japanese army. As it had happened in Spain, Japan made full use of air raids against civilian population, especially in the cities that were soon to be invaded by the Japanese troops. However, while victims could be numbered by thousands, raising medical concerns about the capacity of dealing with emergencies, these attacks did not represent a decisive turning point in the military strategy as the cities kept resisting until they were invaded by regular troops. Meanwhile, the impact of air raids upon the civilian population were widely shown in the international media causing a global reaction of repulse that fueled the activities and membership of the RUP/IPC.

The organization of RUP/IPC consisted of national committees, reaching 43 countries in its peak, institutional affiliates (the organization acted as an umbrella for other associations) and an international committee that was led by French and British affiliates, and a council that was more diverse.[19] In February 1936, the Spanish committee of the RUP/IPC was established in the Ateneo of Madrid, with the name Unión Universal por la Paz (hereafter UUP). It was composed of high-level politicians and intellectuals: the future head of state Manuel Azaña, the socialist Julio Álvarez del Vayo (future minister), the lawyer Ángel Ossorio y Gallardo (that would be the first director of UUP before being assigned as Spanish ambassador in Paris), the physician Teófilo Hernando, and the well-known poet Antonio Machado. They signed the first manifesto of the organization in Spain which was not made public immediately because the

[19] "Une tâche pour chaque homme de bonne volonté. Aidez le Rassemblement Universel pour la Paix", in Bibliothèque La Contemporaine, Jules Prudhommeaux archives (available at archives.org), Mouvements pacifistes. Brochures, Rassemblement universel pour la paix. 1924–1939, F delta res 717 (02) (03), p. 26.

"ambient of intense passion" due to the elections that would give the victory to the popular front, in the last months of peace. The manifesto received the support of poet Federico García Lorca (who would be killed short afterwards) and referred to the Italo Abyssinian crisis. It also aimed to alerting the Spanish audience about the growing threats of war.[20]

The UUP was preparing a delegation to attend the Universal Congress for Peace in Brussels when Franco led a coup d'état and the civil war ensued. Before being appointed as ambassador in Paris, the head of the UUP, the catholic lawyer Ángel Ossorio y Gallardo traveled to Geneva with the Spanish delegation to participate in the debate about Spain in the Society of Nations. There, he was frustrated by the aggressive passivity of the Western powers that were impeding the Republican government to purchase weapons in the open market, while Germany and Italy were actively supporting the rebellion (Viñas, 2006; Jorge, 2016). In November, Madrid was threatened by the troops of Franco that stood in the outskirts of the city and air raids adopted a new face: daily bombings became a common practice. The images of fallen buildings and the corpses of children and women were reported extensively. However, quite surprisingly, the population of Madrid got used to it, scenes of panic were rarely seen, and the city ultimately resisted, against all previsions (Solé i Sabaté & Villarroya, 2003).

In January 1937, after two months of frequent air raids in Madrid, an international conference was organized in Paris to centralize the international medical aid to the Spanish Republic, especially considering the sanitary emergency of air raids not only against cities, but also against hospitals and ambulances.[21] The doctors from eleven countries committed to create national organizations and launch political campaigns to send medical aid to Republican Spain, recruiting volunteers and purchasing medical supplies and ambulances. The international medical aid that was transferred to Republican Spain during these years was important and has been studied elsewhere (Brasó Broggi, 2022). The conference not only criticized the passivity of the European governments that blocked the capacity of purchase of Spain, under the policy of non-intervention, but also the Red Cross, who tried to follow a strict neutrality, which was ultimately

[20] "Un manifiesto de la Unión Universal por la paz", IISH/RUPA, No. 232, p. 3.

[21] *Aidez l'Espagne. Conference Internationale d'Aide Sanitaire à l'Espagne Republicaine*, 16–17 January 1937, pp. 4–5. I would like to thank Julia Lange for providing me with a copy of this rare manuscript that is held at the Bayerische Staatsbibliothek.

benefiting general Franco and his German and Italian allies, which were supplying the rebel generals without obstacles (Marqués, 2000). Furthermore, in China, the conservative board of China's Red Cross hindered the internationalist initiatives of some of its members, like the prominent doctor Robert Kho-seng Lim and the Medical Relief Corps which played a relevant role in providing medical aid in the first years of the Japanese invasion (Watt, 2014; Chi, 2017; Shi, 2018; Ho, 2017, 2020).

The RUP/IPC appealed to the founding principles of the Red Cross and the "international peace conferences" of the end of the nineteenth century. However, when wars came without formal declaration in Abyssinia, Spain, and China, most of the international body of rules created by the Red Cross and the League of Nations proved obsolete. According to the British ambassador to Tokyo, there was a racial and colonial argument about the fact of not declaring war formally. By not declaring war, the aggressors were denying the status of state governments to the territories under attack, which was an implicit way of recognizing them unable to govern themselves.[22] Nehru also believed that by defending the "collective security" of the League of Nations, the organization was favoring the *status quo*, while he thought that peace could only be achieved after the fall of imperialism, as imperialism was the "negation of peace" (Louro, 2018).

Air raids blurred the traditional division between the front and the rearguard. The fact that anybody could be targeted challenged the traditional ways in which the governments provided with medical aid toward the victims of war: women and children were also at the frontline and deserved to be attended in the same manner as conventional soldiers (Grayzel, 2012). This idea was emphasized not only by the Communist rhetoric of the people's war and the united popular fronts but also by the Nationalist anti-imperialist narratives that saw in air raids an argument for calling to the nation's collective rise (a national uprising) against a foreign aggression. The international character of the RUP/IPC, mainly composed of cosmopolitan doctors and politicians, did not deter them to engage in these nationalistic discussions, emphasizing the growing detachment of the colonial population, which could be aggravated with air raids, as well

[22] R. Craigie, "Description of existing relations between China and Japan as comprising an "incident" and the State of war", Tokyo, 25 January 1938, in The National Archives, FO 371/22108, pp. 12–18.

as the traditional passivity of colonial governments concerning issues of public health or well-being (Amrith, 2006).

From May to November 1937, the city of Paris was the showcase of the World Expo where countries and international organizations could display their most relevant cultural and scientific features. The Guernica painting of Pablo Picasso, showing the consequences of German air raid against the Basque town of Guernica, was exhibited for the first time. The bombing took place in April 26 totally devastating the Basque town of Guernica, an important spot in the history of Basque nationalism, with hundreds of victims. Since then, the Guernica painting, full of various symbolisms, has become a flag of the civilian victims of war. It appeared in a context where the Republican propaganda showed photos of Spanish women and children attacked by German and Italian foreign pilots. Meanwhile, the RUP/IPC kept its international stance and set a pavilion in the World Expo on its own, the Peace Pavilion, that intended to transcend national narratives and focus on the need of enhancing international peace (Namer, 1981; Ren, 2018). However, because the expo was nation-based, it was located outside the main venue (Farrugia, 2011).

When Japan started the occupation of China in July 1937, Spain was commemorating the first anniversary of its civil war. Despite the continuous air raids, the cities of Madrid, Barcelona, and Valencia were resisting, but the Republican government was struggling in the Battle of Brunete, in the vicinities of Madrid. Both wars became the center of international attention, not only because they coincided chronologically but also because they shared some common elements: the generalization of air raids, the non-intervention of the Western countries, the establishment of government alliances that integrated communist and noncommunist members (the so-called popular front in Spain and united front in China), the interest of the Soviet Union in helping these governments, and the impact of the international leftist media to put Spain and China as the vanguard of a global war against fascism. According to this narrative, as Madrid was resisting, so Wuhan did, and both cities symbolized their national spirit of anti-fascist resistance (Borao, 2017).

In January 1938, the Chinese committee of RUP/IPC was created in Wuhan with the name of International Anti-Aggression Movement (hereafter AAM).[23] The AAM set up its headquarters in Wuhan until the city finally fell over by the Japanese and had to move to Chongqing, with the

[23] In Chinese, *Guoji fanqinlüe yundong*.

rest of the Chinese nationalist government. The AAM had 150 institutional affiliates and a permanent committee of 23 members, under the high-profile presidency of the brothers T. V. Soong (Song Ziwen), Nationalist minister, and his sister and widow of Sun Yatsen, Soong Ching-ling (Song Qingling). The AAM also set local delegations in Guilin, Kunming, Guiyang, and permanent international delegations in Geneva, London, and Paris to coordinate the international propaganda activities in favor of China.[24]

As it happened to Spain, the RUP/IPC sent a special correspondent to China. In November 1937, the vice-president of the international committee of RUP/IPC, the protestant pastor Jules Jézéquiel, traveled to Republican Spain to report about the conditions of the country. He visited Barcelona, Valencia, and Madrid and depicted the stoic attitude of the urban dwellers that were carrying out "business as usual" despite the continuous threat of air raids. Jézéquiel also enquired about the religious practices in Republican Spain (he attended several protestant masses) and the conditions in the front, especially in the outskirts of Madrid. He gave a positive image of the Republican government.[25]

The British journalist Vernon Bartlet, having acquired experience as a war reporter in Spain, traveled to Hong Kong, Shanghai, and Wuhan during two months in the beginning of 1938. He was received by the general Chiang Kai-shek, which is indicative of the importance of the organization in these years. Bartlett was in Guangzhou during one of the worst air raids and saw the destruction of the district of Zhabei, the Shanghainese town that stood just beside the International Concession of Shanghai.[26] Japanese airplanes would not attack the neutral international concessions of Shanghai, while bombing the Chinese neighborhoods. Bartlet was shocked by the contrast with the apparent normal life of the colonial concessions, the main economic hub of China and where the major British colonial interest stood, where the people would sit in a bar terrace and look to the

[24] See *Rapport du Comité national chinois du R.U.P., présenté au Conseil général du Rassemblement universel pour la paix, 31 décembre 1938*. Report submitted by the Chinese national committee of the I.P.C. to the General council of the International peace Campaign, 31 December 1938. Available at Bibliothèque Nationale de France, (January 25, 2022): http://catalogue.bnf.fr/ark:/12148/cb339403647

[25] J. Jézéquiel, "Impressions d'Espagne. 15 jours en Espagne Républicaine" (11–25 November), IISH/RUPA, No. 143.

[26] Vernon Bartlett, "Report on the situation of China", no date (1938), in IISH/RUPA, No. 171.

war developing at the other side of the river. He concluded that the British Empire was not understanding the new era of massive violence and that a global war was, thus, imminent.[27]

Following the creation of the UUP, the RUP/IPC organized a major event in London, an international conference called "Save China, save Peace" (February 12–13, 1938). It called for a massive campaign against air raids in China that would collect medical aid and for a boycott of Japanese products.[28] The Comintern agent Louis Dolivet, as international secretary of RUP/IPC, was behind the campaign that copied the modus operandi of Münzenberg, especially the anti-imperialist campaigns of 1925 that denounced the "atrocities of imperialism" in China, with a massive boycott of British and Japanese goods and huge demonstrations in Shanghai. Furthermore, a confidential document of the Comintern from October 1937 called to all working-class parties to undertake a "tremendously powerful campaign the center of which will be concentrated on the defense of Republican Spain and China."[29]

The conference in London of February 1938 gathered 800 members of 21 countries that demanded for their respective governments to block all trade with Japan. It included a specific commission on medical aid for China that was led by the British Quaker reformist Margery Fry (also one of the first women to become magistrate in the United Kingdom). The commission brought the following conclusions: to organize funding activities internationally to send medical aid to China; to focus these activities on concrete medical needs, such as ambulances or medical supplies for hospitals that had suffered air raids; and to establish an international cooperation committee to supervise the activities that were being carried out nationally.[30]

Just after the conference, the RUP/IPC sent two more journalists in China as delegates of the organization: the French Pierre Scize, editor of *Paris Soir*, and Edgar E. Mowrer, from the *Chicago Daily News*. Again, Scize could interview Chiang Kai-shek and witness an air raid in Guangzhou, the only relevant city of coastal China that stood free from

[27] Vernon Bartlett, "La sécurité collective, seule, peut écarter la guerre", *Journal des Nations*, 10–11 April, 1938, p. 3; in IISH/RUPA, No. 171.

[28] Louis Dolivet to the members of RUP/IPC, 17 December, 1937, in IISH/RUPA, No. 162, pp. 2–3.

[29] "On aid for the Chinese and Spanish people", October 1, 1937, RGASPI, 495/18/1225.

[30] Résolution de la commission pour l'aide à la Chine, Conference Mondiale pour l'Aide à la Chine et le boycott du Japon, Resolutions, IISH/RUPA, No. 168, p. 4.

the Japanese occupiers.[31] The comparisons between Guangzhou and Barcelona were common in these reports that were published in the international press. Thus, the extreme violence of these air raids in Spanish and Chinese cities was very present in June 1938, when the mayor of Guangzhou sent the message of help and the RUP/IPC decided to organize an international conference exclusively focusing on this topic.[32]

The International Conference Against Air Raids

Following the call of the mayor of Guangzhou, the secretary general of RUP/IPC, Pierre Cot, wrote personally to the mayors of French cities calling for a public condemnation of the bombings. The campaign against air raids in open cities created a massive movement of support among city governments that countered the official non-intervention of central governments. In France, more than 200 cities and towns answered to the call of Pierre Cot; in the United States, more than 60 cities gave their support (among them, New York, Los Angeles, San Francisco, Detroit or New Orleans, etc.); in Great Britain, the London City Council and other major cities reacted (Manchester, Bristol, Glasgow); and other urban governments that gave their support to the campaign were Stockholm, Oslo, Prague, Pilsen, Geneva, Lausanne, Brussels, and Antwerp.[33] Despite the nationalistic propaganda, the RUP/IPC built a kind of parallel urban diplomacy that was led by leftist intellectuals and politicians that aimed at countering the passivity of the moderate central governments of France, Britain, and the United States.[34]

These cities showed their solidarity toward Guangzhou, but they also had a practical interest in the subject of air raids, as some suspected that they soon might become targets. British city boroughs governed by the Labour Party learned the construction of air-raid shelters following the Spanish experience of the active participation of the population (the "people's war" as it was known), thanks to the visits of experts like the

[31] Pierre Scize, "Interview avec Chang Kai Chek", Hong Kong, 29 April, 1938,; Pierre Scize, "Quand les japonais bombardent Canton"; in IISH/RUPA, No. 174.

[32] Pierre Scize, "Retour de Chine", in IISH/RUPA, No. 174.

[33] "A new threat to mankind: aerial terrorism. The appeal of the mayor of Canton and the protests of municipalities around the world", August, 1, 1938, in IISH/RUPA, No. 176, pp. 15–20.

[34] Reports of Consul General of Britain in Canton, General Blunt, June 4, 1938, The National Archives, FO 371/22039, pp. 4–14.

heterodox scientist J. B. S. Haldane (Haapamaki, 2009). Haldane was a member of an influent leftist group based at Cambridge University that became known as the Cambridge Scientists Anti-War Group, whose members were committed in defending the causes of the Spanish Republic and the Chinese war against Japan. Among them stood the brilliant scientist Joseph Needham who became known for his investigations about the history of science in China, challenging the mainstream theory that science was a uniquely European phenomenon (Chakrabarti, 2004). Between 1938 and 1939, the findings of Haldane in Spain brought about a general discussion in Britain about the necessity of building public shelters for air raids (Haapamaki, 2009). Meanwhile, the Chinese ambassador in London was instructed to watch and study how the Spanish cities had managed to protect themselves against air raids.[35]

In the spring of 1938, the campaign against air raids of RUP/IPC gained momentum. The images of dead bodies of children after the bombardments of Barcelona and Guangzhou aroused international protests and demonstrations. In May, the International Committee of the Red Cross had made a public statement condemning the bombardment of cities.[36] However, the Red Cross did not appeal to any government, following a strict neutrality that reminded the non-intervention policies of the Western countries. Even though the Chinese and Spanish governments also used air raids against enemy cities, their impact was limited to military spots (occasionally hitting civilian targets) and exhaustive bombings to cities were not practiced. Thus, the clear positioning of the RUP/IPC against Franco, Italy, Germany, and Japan, in contrast to the elusive discourses of other international organizations and some central governments, favored its international expansion, reaching a global audience.

The conference was celebrated in Paris in July 23–24, which gathered more than 1000 attendants from 34 countries making it as one of the largest events organized by the RUP/IPC. The sessions were chaired by both the president of the Spanish parliament, Mr. Martínez Barrio, and the president of the National Academy of China, Mr. Li Hua-yu. However, this conference also made visible the differences and contradictions of the organization. The Spanish conflict had created a split between the liberal

[35] Reports of Consul General of Britain in Canton, General Blunt, June 4, 1938, The National Archives, FO 371/22039, p. 4.

[36] "La 348 Circulaire aux comitès centraux des Societés nationales de la Croix-Rouge, en date du 30 mai 1938, donne son sentiment", quoted in Marqués, pp. 412–13.

and the leftist members of the organization (Farrugia, 2011). Therefore, the tone of the opening address made by Pierre Cot and Lord Cecil tried to be diplomatic and balanced, emphasizing the non-political character of the organization while criticizing the lack of commitment of the great powers. This bias continued in the following discussion, with rather abstract references to the need to fight for peace and the barbarian character of air raids against open cities. At a more concrete level, however, Martínez Barrio explained the cases of four Spanish cities (Badalona, Blanes, Alicante, and Granollers) that were bombed despite lacking any military interest, while Nehru denounced the British bombing of the North-West parts of India and Li Yu-ying referred to his recent visit to Wuhan.[37] The diplomatic tone of the British and French speakers was confronted by the cold exposure of facts from the affected countries.

In the conference, Nehru informed that a medical mission supported by the Indian National Congress would be destined for China. The mission was led by Dr. Menhanlal Atal, which was a part of the Indian delegation in the conference. From Uttar Pradesh, Dr. Atal had gained experience as a war doctor in Spain and led a team of five Indian doctors (besides himself, M. Cholkar, Dwarkanath Kotnis, B. K. Basu, and Debesh Mukherjee) that would leave Britain in August, reaching the Communist capital of Yan'an in February 1939 (Jiang, 1982; Framke, 2016).

In August, Nehru decided to visit China and was received at Chongqing airport by Dr. Mukherjee, from the Indian Congress Medical Unit in China (Nehru, 1940). The impact of air raids in Spanish and Chinese cities left a visible imprint in Nehru's vision of the world affairs, as he detailed in a book published in 1940, titled *China, Spain and the War. Essays and Writings*. However, the Indian leader was also a supporter of air traveling that had made possible his visits to wartime Spain and China: "After long hesitation I decided to go to China. I did so because far as she is, air travel has brought her near to us within two or three days reach" (Nehru, 1940). Nehru's positioning against air raids was not against technology, but against imperialism.

If the international medical aid to Spain was coordinated from Paris, the campaign to send medical aid to China was based in London even though both flows of aid crisscrossed internationally, favoring the spillover of medical innovations. These flows of medical aid were led by several organizations linked to RUP/IPC, like the China Campaign Committee,

[37] *Au secours des victimes de l'agression*, op. cit.

from whom the Foreign Office suspected that were led by Communist and Indian nationalist members who could endanger the British colonial interests.[38] The China Campaign Committee appeared in September 1937 under the patronage of Lord Cecil and the support of the Chinese embassy in London. Like the RUP/IPC, it integrated a variegated range of leftist intellectuals, labour politicians, former volunteers of the Spanish International Brigades, Communist party members (Ben Bradley, Arthur Clegg), the Indian editor and the nationalist leader V. K. Krishna Menon, who was also a member of the international council of RUP/IPC; and Chinese sojourners in London like Pat Koo, the son of diplomat Wellington Koo; and Shelley Wang or Wang Lixi (Clegg, 1989; Buchanan, 2012).

The Liberal Party member of the British Parliament, Richard Acland, answered Nehru's accusations claiming that the public opinion of Great Britain was ashamed by the practice or air raids in India. He also announced that the British delegation was funding another medical mission in China that would build an International Peace Hospital.[39] When the conference against air raids was taking place in Paris, Norman Bethune, after having created a blood transfusion center in Madrid, was already working as a field doctor in the remote area of Jin-Cha-Ji (in the mountains Wutai, in the border area of Shanxi-Chahar-Hebei), under the Eight Route Army led by the Communist generals Nie Rongzhen, Zhu De, and, ultimately, by Mao Zedong (Allan & Gordon, 1954; Stewart & Stewart, 2011). In the summer of 1938, the British Committee of the RUP/IPC and other leftist organizations sent funds for establishing the International Peace Hospital in the Communist most isolated regions of China, a hospital that would be led by Dr. Bethune (Clegg, 1989). However, Dr. Bethune died in November 1939 of septicemia and it was the Indian medical mission, led by Dr. Kotnis who substituted Atal, who led the International Peace Hospital that would be renamed Bethune International Peace Hospital.

Meanwhile, as the prospect of massive air raids became a popular theme of discussion, Dr. Josep Trueta and Dr. Duran Jordà migrated to London and released a report of how medical doctors and the urban hospital systems should deal with air raids, following their experiences in Barcelona. In March 1939, the report about "medical aspects of air raids casualties in Barcelona" gathered the experience of 10,000 casualties of a total of 350

[38] "Activities of the China Campaign Committee". 25 de abril de 1940, The National Archives, Foreign Office, FO 371/24701, pp. 462–468.

[39] *Au secours des victimes de l'agression*, p. 73.

air raids that took place in the Catalan city between 1936 and 1939: 17 per cent of them died before even reaching the hospital.[40] Thanks to this experience the city of London was better prepared for the dangers that would soon have to face. In April 1939, the US doctor John Scudder published a joint article with references to Duran Jordà that became a standard in blood transfusion practices for Great Britain and the United States during the Second World War. Furthermore, Scudder went to China in 1944 and installed the first blood bank center in the city of Kunming (Soon, 2020). In June 1940, in the midst of the blitz against London, Churchill would evoke the capacity of resistance of Barcelona (Solé i Sabaté & Villarroya, 2003).

The RUP/IPC did not intervene directly in these medical transfer operations, but supported them through organized funding events, conferences, debates, and propaganda materials. The conference of air raids was a platform for parallel urban and non-governmental diplomacy that challenged the official non-intervention policies of most of the countries. The proceedings included letters and messages sent by city mayors to the call of the mayor of Guangzhou. Expressions of solidarity came from hundreds of cities and towns from France, Great Britain, United States, Mexico, Sweden, Norway, Switzerland, Czechoslovakia, New Zealand, Belgium, India, Egypt, Jamaica, Bolivia, Ceylon, South Africa, and Palestine.[41] However, when looking at the concrete political propositions that came at the end of the journey, their impact was rather limited. The conference concluded that Spain and China should be allowed to purchase weapons for their self-defense to receive financial and medical help and to facilitate the evacuation of civilian population to secure zones. The RUP/IPC also committed to organize an international campaign to ban air raids, to launch boycotts against Japan, and to send medical aid and basic goods to both countries. Finally, there was an open letter asking for help to the US President Franklin D. Roosevelt.[42]

The press of Barcelona, one of the cities that had been suffering the most, was not very enthusiastic regarding the results of the conference. They doubted that these declarations of good intentions would mean a

[40] Josep Trueta Raspall, "Opinions formed from actual experience of Aerial bombardment of Barcelona", in The National Archives, FD 1/5372, pp. 35–55.

[41] "A new threat to mankind: aerial terrorism. The appeal of the mayor of Canton and the protests of municipalities around the world", August, 1, 1938, in IISH/RUPA, No. 176, pp. 15–20.

[42] "Resolutions", in *Au secours des victimes de l'agression*, pp. 92–93.

change in the situation of the city, where the fiercest battle of the Spanish Civil War (the Ebro battle) was taking place, just 150 kilometers away.[43] The leadership of France and Great Britain in RUP/IPC was seen with suspicion, as both countries had led the non-intervention policies that had been undermining the capacity of the Republican government.

The crisis of the RUP/IPC became evident just after the conference with the end of the popular front government of France (who was the main financer of RUP/IPC) and the German aggression of Czechoslovakia, a country that had played an important role in the deployment of activities of RUP/IPC. The Munich agreement, some months after the congress, drew a heavy blow to the organization's stance for "active pacifism" in Europe.

Another factor of distress came with the petition of the organization to the French and British colonial governments to open their frontiers in the Chinese border (the British in Burma and the French in Indochina) to allow medical aid to be transported to China.[44] The imperial interests of Great Britain and France were confronted in Asia, dividing the British and French branches of RUP/IPC and their colonial counterparts. In 1939, this gap widened when the Indian committee of RUP/IPC announced a conference in Allahabad—the hometown of Nehru and where the main office of the Indian section of the RUP/IPC stood—inviting the Chinese delegation and the Chinese foreign ministry T. V. Soong (Song Ziwen) (Louro, 2018).[45] The British section was not invited.

Meanwhile the Chinese press had a more positive approach about the campaigns organized by RUP/IPC, especially regarding the importance of boycotting Japanese goods, a campaign that was growing strong in the United States.[46] However, while conferences and demonstrations went by in Paris and London, the cities of Wuhan, Guangzhou, Barcelona, and Madrid fell to the enemy after continuous air raids, without a major change in the policy of non-intervention. In January 1939, a delegation of 700 members and sympathizers of RUP/IPC aimed to meet with the British

[43] Fabra, "Se inauguró la conferencia internacional sobre los bombardeos de ciudades abiertas", *La Vanguardia*, July 24, 1938, p. 11.

[44] Lord Halifax to British Executive Committee, 15 November 1938, The National Archives, Foreign Office, FO 371/22075, p. 10.

[45] Proposed Chinese delegation for International Peace Campaign Conference at Allahabad, The National Archives, Foreign Office, FO 371/23535, pp. 321–28.

[46] Guoji fanqinlüe yundong dahui (Conference of the Anti-Aggression movement), *Minyi (Hankou)*, 1938, Vol. 10, pp. 9–10.

government just before the visit of Chamberlain to Mussolini to denounce the British connivance toward the Italian fascist leader. The delegation was not even received.[47] Conferences and demonstrations in favor of international peace continued to be celebrated in Britain and France by the RUP/IPC until the summer of 1939.[48]

Conclusions

The air raids against open cities stood in the middle of the heated discussions around imperialism, fascism, and nationalism that dominated the public debate in the years before the Second World War. The debate delved with the possibility that pacifism could give support to military actions when countries and regions suffered from the aggression of others, especially of what Nehru called "intense nationalism." Other countries like Czechoslovakia, China, or Spain also claimed their right to defend themselves, but the RUP/IPC did not establish the limitations of nationalism in the use of force and the right of self-defense especially when wars were not formally declared as it often happened in colonial conflicts. Therefore, the message of RUP/IPC against the use of air raids, although good intentioned, was somehow paradoxical and short-lived. After the invasion of Poland in September 1939, air raids became widespread, not only in Warsaw, London, Rotterdam, Minsk, Rangoon, Singapore, and Manila but also in Dresden, Hamburg, Berlin, Tokyo, Hiroshima, and Nagasaki.

However, other proposals of the organization were more enduring. The RUP/IPC perceived the inefficiency of the League of Nations and the non-intervention policies of France and Great Britain as a clear sign of the diminishing power of Europe and the rise of the Soviet Union and the

[47] RUP/IPC, "Le voyage à Rome. Députation nationale à Mr. Chamberlain", Geneva, January 26, 1939, in Bibliothèque La Contemporaine, Jules Prudhommeaux archives, Mouvements pacifistes. Brochures, Rassemblement universel pour la paix. 1924–1939, F delta res 717 (02) (03), p. 34.

[48] Louis Dolivet a Prudhommeaux, June 5, 1939; and Jammy Schmidt, 1 July 1939, Paris, Bibliothèque La Contemporaine, Jules Prudhommeaux archives, Mouvements pacifistes. Brochures, Rassemblement universel pour la paix. 1924–1939, F delta res 717 (02) (03), p. 34.

RUP/IPC, "Le voyage à Rome. Députation nationale à Mr. Chamberlain", Geneva, January 26, 1939, in Jules Prudhommeaux archives, Mouvements pacifistes. Brochures, Rassemblement universel pour la paix. 1924–1939, F delta res 717 (02) (03), pp. 5–6.

United States as major global powers. The organization also acknowledged the emancipation of the regions that had been colonized by European countries or that had suffered from imperialist or fascist aggressions. Air raids that had been used by both the French and British Empire before the wars of Spain and China were symbols of imperialism and its inner violence, showing the lack of care of the local civilian population (symbolized by the fact that women and children were often victims of air raids). While it was possible that air raids and air policing extended the means of the British Empire to prolong its existence, they also enforced national resistance and active opposition among the civilian population.

Even though some of the political discourse of the organization and their statements may seem naïve when looking to the past of 1938, the RUP/IPC was a notable platform of networking between leaders and intellectuals of different backgrounds, allowing knowledge transfers and mediations. The medical knowledge about how to deal with air-raid victims and the transfer of doctors was a valuable outcome of these meetings. It was also a place where international leaders could discuss issues of international relations on an equal basis. These were perhaps the most enduring legacies of the RUP/IPC.

References

Archives

Amsterdam, International Institute of Social History (IISH), Rassemblement Universel pour la Paix Archives (RUPA).
Archive of the League of Nations.
Archives Nationales (France), Archives Sureté Nationale.
Bayerische Staatsbibliothek.
Bibliothèque La Contemporaine, Jules Prudhommeaux Archives (Internet Archives).
Bibliothèque Nationale de France.
New York University Archives, The Taminent Library & Robert F. Wagner Labor Archives, Archives of the Abraham Lincoln Brigade Archives (ALBA), Fredericka Martin Papers.
Russian State Archive of Socio-Political History (RGASPI), Fond 495, Executive Committee of the Communist International (ECCI).
The National Archives (Kew Gardens, London), Foreign Office Archives.

Other Sources

Allan, T., & Gordon, S. (1954/2004). *The Scalpel, the Sword. The Story of Dr. Norman Bethune*. Foreign Language Press.

Amrith, S. (2006). *Decolonizing International Health. India and South-East Asia, 1930–1965*. Palgrave Macmillan.

Arañó, L., & Capdevila, M. (2019). *Topografía de la destrucció. Els bombardeigs de Barcelona durant la Guerra Civil (1936–1939)*. Ajuntament de Barcelona.

Bethune, N. (1937). *The Crime on the Road of Malaga-Almería*. Iberia.

Borao Mateo, J. E. (2017). *Las miradas entre España y China. Un siglo de relaciones entre los dos países (1864–1973)*. Miraguano Ediciones.

Brasó Broggi, C. (2022). *Los médicos errantes. De la guerra civil y la revolución China a la guerra fría*. Crítica.

Broggi, M. (1997). Sobre Frederic Duran i Jordà. *Gimbernat, 27*, 185–191.

Buchanan, T. (2012). *East Wind: China and the British Left, 1925–1976*. Oxford University Press.

Chakrabarti, P. (2004). *Western Science in Modern India. Metropolitan Methods, Colonial Practices*. Permanent Black.

Chi, Z. (2017). *Zhongguo hongshizi yundongshi ziliaobian*. Hefei Gongye Daxue Chubanshe.

Clegg, A. (1989). *Aid China 1937–1949. A Memoir of a Forgotten Campaign*. New World Press.

Farré, S. (2007). Mobiliser, unir, sauver: l'aide humanitaire durant la guerre civile espagnole. *Cahiers d'historie du mouvement ouvrier, 23*, 68–84. https://doi.org/10.5169/seals-520299

Farrugia, P. (2011). Mésentente cordiale: Anglo-French Collaboration in the Rassemblement Universel pour la Paix. *Synergies RUI, 4*, 105–116.

Framke, M. (2016). Political Humanitarianism in the 1930s: Indian Aid for Republican Spain. *European Review of History: Revue européenne d'histoire, 23*(1–2), 63–81. https://doi.org/10.1080/13507486.2015.1117421

Gallego Vila, L. (2017). Els bombardeigs de Barcelona durant la guerra civil: historiografia i memorialització. *Índice Histórico Español, 130*, 13–37.

Grayzel, S. R. (2012). *At Home and Under Fire. Air Raids and Culture in Britain from the Great War to the Blitz*. Cambridge University Press.

Grífols i Espés, J. (1997). *Frederic Duran i Jordà. Un mètode, una època*. Hemo-Institut Grífols.

Haapamaki, M. A. (2009). *The British Left Wing and Air Raid Precautions, 1918–1939*. PhD Thesis, McMaster University. Thesis Available at: https://macsphere.mcmaster.ca/handle/11375/17304. Last 1 Feb 2022.

Ho, B.-L. (Ed.). (2017). *Lin Kesheng minguo yixue shishang diyiren/A Collection of Essays in Memory of Dr. Robert Kho-Seng Lim (1897–1969)*. Zhonghua kejishi xuehui.

Ho, B.-L. (2020). *Lin Kesheng, Zhuisi lunwenlu (1891–1969)/The Founder of Chinese Physiology and Military Medicine.* Guojia Tushuguan chubanshe.
Holman, B. (2014). *The Next War in the Air. Britain's Fair of the Bomber, 1908–1941.* Routledge.
Hutchinson, J. F. (1996). *Champions of charity. War and the rise of the Red Cross.* Westview.
Jiang, Y. (1982). In Loving Memory of India's Heroic Dr. Kotnis. *Beijing Review, 25*, 27–28.
Jorge, D. (2016). *Inseguridad colectiva. La Sociedad de Naciones, la guerra en España y el fin de la paz mundial.* Tirant Editorial Humanidades.
Louro, M. L. (2018). *Comrades Against Imperialism Nehru, India, and Interwar Internationalism.* Cambridge University Press.
Marqués, P. (2000). *La croix-rouge pendant la guerre d'Espagne (1936–1939). Les missionnaires de l'humanitaire.* L'Harmattan.
Mazuy, M. R. (1993). Le Rassemblement Universel pour la Paix (1931–1939): une organisation de masse? *Matériaux pour l'histoire de notre temps, 30*, 40–44.
Moral Torres, J. (1986). El método español en el tratamiento de las heridas de guerra. In *Los médicos y la medicina en la guerra civil española* (pp. 159–166). Beecham.
Namer, G. (1981). Les imaginaires dans l'exposition de 1937. *Cahiers Internationaux de Sociologie, 70*, 35–62.
Nehru, J. (1940). *China, Spain and the War. Essays and Writings.* Kistabistan.
Neill, D. (2012). *Networks in Tropical Medicine: Internationalism, Colonialism, and the Rise of a Medical Specialty, 1890–1930.* Stanford University Press.
Olier, F. (2002). Les auto-chirs (1914–1918). Genèse d'une épopée. *Medécine et armées, 30*(299), 320.
Omissi, D. E. (1990). *Air Power and Colonial Control. The Royal Air Force, 1919–1939.* Manchester University Press.
Palfreeman, L. (2015). *Spain Bleeds. The Development of Battlefield Blood Transfusion During the Civil War.* Sussex Academic Press.
Petersson, F. (2013). *We Are Neither Visionaries Nor Utopian Dreamers. Willi Münzenberg, the League Against Imperialism, and the Comintern, 1925–1933.* Doctoral Thesis, General History, Division for Arts, Education and Theology, History Department, Åbo Akademi University.
Ren, K. (2018). The International Peace Campaign, China, and Transnational Activism at the Outset of World War II. In C. P. Peterson, W. M. Knoblauch, & M. Loadenthal (Eds.), *The Routledge History of World Peace Since 1750* (pp. 359–370). Routledge.
Rodrigo, A. (2009). *Doctor Trueta. Héroe anónimo de dos guerras.* Editorial Base.
Satia, P. (2006). The Defense of Inhumanity: Air Control and the British Idea of Arabia. *American Historical Review, 111*(1), 16–51. https://doi.org/10.1086/ahr.111.1.16

Schauff, F. (2006). *La victoria frustrada. La Unión Soviética, la Internacional Comunista y la guerra civil española* (G. Sans Mora, Trans.). Debate.

Shi, Y. (2018). *Lin Kesheng yu minguo xiandai yixue de fazhan*. Guojia tushuguan chubanshe.

Solé i Sabaté, J. M., & Villarroya, J. (2003). *España en llamas. La guerra civil desde el aire*. Temas de Hoy.

Soon, W. (2020). *Global Medicine in China. A Diasporic History*. Stanford University Press.

Stewart, R., & Stewart, S. (2011). *Phoenix. The life of Norman Bethune*. McGill-Queen's University Press.

Trueta, J. (1939). The Treatment of War Fractures by the Closed Method. *Proceedings of the Royal Society of Medicine, 33*(1), 65–74.

Viñas, Á. (2006). *La soledad de la República. El abandono de las democracias y el viraje hacia la Unión Soviética*. Crítica.

Watt, J. R. (2014). *Saving Lives in Wartime China: How Medical Reformers Built Modern Healthcare Systems Amid War and Epidemics, 1928–1945*. Brill.

Correction to: From Universal Rats to Future Jungle Foci: Actors and Places of Plague in Brazil (1899–1940s)

Matheus Alves Duarte da Silva

Correction to:

Chapter 7 in: M. Capocci, D. Cozzoli (eds.), *Empire, Nation-building, and the Age of Tropical Medicine, 1885–1960*, Medicine and Biomedical Sciences in Modern History, https://doi.org/10.1007/978-3-031-38805-7_7

The chapter "From Universal Rats to Future Jungle Foci: Actors and Places of Plague in Brazil (1899–1940s)", Matheus Alves Duarte da Silva, was originally published without open access. Following the author decision to opt for open access, the copyright of the chapter changed on 01 February 2024 to © The Author(s) 2024 and the chapter is now distributed under the terms of the Creative Commons Attribution 4.0

The updated version of this chapter can be found at
https://doi.org/10.1007/978-3-031-38805-7_7

M. Capocci, D. Cozzoli (eds.), *Empire, Nation-building, and the Age of Tropical Medicine, 1885–1960*, Medicine and Biomedical Sciences in Modern History,
https://doi.org/10.1007/978-3-031-38805-7_9

Index[1]

A
AAM, *see* Anti-Aggression Movement
About, I., 4
Abyssinia, 182, 192
Academy of Medical Science of Soviet Union, 111
Acclimatization, 35, 147
Accurso, Salvatore, 17, 147, 153, 154
Ackernecht, E. H., 3, 7
Acokanthera, 150
ACPII, *see* Advisory Committee for the Plague Investigation in India
Adler, S., 106–108, 106n9, 107n13, 117, 121, 123–125
Advisory Committee for the Plague Investigation in India (ACPII), 164
Adwa, battle of, 141, 154
Aedes aegypti, 112, 175
Africa, 2–4, 12–14, 18, 19, 30n1, 31–33, 35, 39–49, 57–59, 83, 84n8, 93, 117, 119, 125, 141, 145, 153, 154, 169
Aggeu Magalhães Research Center, 112
Agouti, *see Dosyprocta azarae*
Aguilar, Saraiva de, 44
AIDS, 127
Air raids, 182–203
Alencar, Joaquim Eduardo de, 110, 110n16, 111, 120n23
Alexandre, Valentim Velho Brasil, 30n1
Alexandria, 19
Ali, Myles, 96
Alicante, 198
Allahabad, 201, 201n45
Allan, Ted, 199

[1] Note: Page numbers followed by 'n' refer to notes.

M. Capocci, D. Cozzoli (eds.), *Empire, Nation-building, and the Age of Tropical Medicine, 1885–1960*, Medicine and Biomedical Sciences in Modern History,
https://doi.org/10.1007/978-3-031-38805-7

Alliance for Progress, 114, 115
Alma-Ata, Declaration of, 15
Álvarez del Vayo, Julio, 185, 190
Alves, Rodrigues, 165
Amaral, Isabel, 18, 30, 32–34, 38, 40–42, 44, 46, 102, 146, 168
Amazon, 105, 122, 125, 150, 151, 152n7, 162, 163, 174–176
Ambaca, 32
Amebiasis, 116
American Civil War, 8
American Confederacy, 57
Amrith, Sunil, 185, 193
Amsterdam, 55, 67, 182n1
Anderson, B., 17
Anderson, Warwick, 14, 17
Andes, 122
Andrade, Nuno de, 165
Angola, 9, 30n1, 31–33, 40, 40n6, 41, 44–46, 48, 50, 169
Annali di Medicina Navale, 148
Anopheles, 13, 76, 81, 111, 118
Anopheles bellator, 119
Anopheles gambiae, 110
Anti-Aggression Movement (AAM), 193, 194, 201n46
Antiimperialist league, 188
Antunes, 109
Antwerp, 196
Aragão, Henrique de B. R, 106, 107n11
Arañó, Laia, 182
Argentina, 8, 10, 16, 102n1, 120n23, 164
Argentine Society of Northern Pathology (SAPRN), 106
Arias, Jorge, 125
Arnold, D., 76
Aruangua, 43
Asia, 8, 12, 19, 33, 58, 63, 64, 103, 107, 117, 125, 153, 165, 169, 201
Assab Bay, 140
Assam, 107
Association for Tropical Diseases, 6
Assumpção, Lucas Virgilio de, 175
Astrakhan, 168
Attal, Menhanlal, 198
Australia, 8, 57
Awooner-Renner, William, 80
Azaña, Manuel, 185, 190

B

Baccari, Eduardo, 153
Badalona, 198
Bahia, 102, 112, 173
Balduino, Candido, 149
Baldwin, P., 4, 7, 16
Ballinger, Pamela, 143
Bangura, Joseph, 79n3
Barbeitas, Mady Malheiros, 105n5
Barcelona, 181–184, 184n6, 189, 193, 194, 196, 197, 199–201
Baroni, Eduardo, 150, 151
Barrell, John, 62, 63
Barrett, Toby Vincent, 125
Barretto, M. P., 121
Barrio, Martínez, 197, 198
Bartlett, Vernon, 194
Bartonellosis, 118
Bashford, A., 15, 57
Bastos Ansart, Manuel, 185
Basu, B. K., 198
Batavia, 61
Bayer, 6, 7, 105n5, 111
Bayly, Christopher, 2
Bay sore, 122
Becker, Howard, 54
Belém, 103n3, 105, 125, 126, 127n32
Belgium, 9, 19, 32, 41, 44, 45, 200
Belize, 124n29
Belli, Carlo M., 147–150, 152
Belo Horizonte, 105n5, 124

Benadir, 17, 147
Benchimol, Jaime Larry, 4, 19, 102, 102n1, 103, 106n8, 108, 112, 125, 125n31, 161, 165, 175
Beneš, Edvard, 185
Beriberi, 41, 42, 148, 149
Berlin, 30n1, 202
Berlin Conference, 2, 32, 50
Bertillon biometric system, 4
Bethune, Norman, 183, 184, 184n5, 199
Bettencourt, Annibal, 33
Bewell, Alan, 63, 64
Bhattacharya, Sanjoy, 5
Biagi Filizola, Francisco, 122
Blaize, Richard, 84, 84n8, 93
Blanes, 198
Blood banks, 20, 183, 200
Bodio, Luigi, 142, 145, 146
Bolivia, 122, 200
Bombarda, Miguel, 35
Bombay, *see* Mumbai
Bondini, Alessandro, 143
Bonelli, Costanza, 142, 142n2, 148, 153, 154
Booth, M., 64
Boston City Hospital, 148
Bottini, Carlo, 148, 149
Bovet, Daniel, 7
Boxer rebellion, 148
Bradley, Ben, 199
Brahmachari, Upendranath, 105n5
Brasó, Carles, 20, 188
Braudel, Fernand, 5
Brazil, 4, 16, 19, 20, 30n1, 101–127, 152n7, 154, 161–176
Brener, Zigman, 120n23
Bressanin, Rodolfo, 150, 151
Brigantaggio, 156n8
Britain, 3, 5–10, 12, 15–17, 19, 30n1, 57, 78, 87–89, 169, 170, 196–198, 202
British Empire, 3, 5, 8, 11, 86, 195, 203
British Guiana, 76
Brock, T., 16
Broggi, Moisés, 184
Bromeliad, 119
Brown, Theodore M., 166
Brumpt, Émile, 120
Brussels, 186, 191, 196
Bruto da Costa, Bernardo, 40–42, 40n6
Bryan, Patrick, 90
Buchanan, Tom, 188, 199
Buenos Aires, 149, 152
Bukovina, 10
Bulhões Maldonado, Angelo, 42
Buonanni, Saverio, 150
Bush, Vannevar, 14

C

Cachin, Marcel, 185
Cadbury, William, 41
Calabria, 156
Calcutta School of Tropical Medicine, 116n18
California, 3, 176
Californian ground squirrel, 168
Calmette, Albert, 163
Câmara Pestana Royal Institute of Bacteriology, 33
Cambournac, Francisco, 47–48, 50
Cambridge Scientists Anti-War Group, 197
Canadian Blood Transfusion Service, 184
Canary Islands, 78, 88
Candau, Marcolino Gomes, 114
Cantlie, James, 57
Canton, 196n33, 200n41
Capdevila, Mireia, 182
Capocci, Mauro, 20, 101, 185

Cappiello, Antonio, 142
Carini, Antonio, 102n1
Carlos I, King of Portugal, 34, 36
Carpi, Leone, 140
Carrion's disease, 118
Castellani, Aldo, 34
Catalonia, 10
Cavalcante, 110
Cavia aperea, 171
Ceará, 109, 110, 110n16, 112, 170n5, 171, 173
Ceará School of Medicine, 110
Cecil, Robert Lord, 185, 186, 198, 199
Central America, 122, 125
Chakravorti, Spivak G., 9
Chagas, Agnes, 106n8
Chagas, Evandro, 103–108, 103n2, 105n7, 107n11, 107n13, 110, 127n32
Chagas disease, 106, 112, 119
Chakrabarti, Pratik, 197
Chamberlain, Joseph, 37n4, 81, 87–89, 93, 202
Chandra, S., 66
Charter of Punta del Este, 114, 115
Chiang Kai-shek, 188, 194, 195
Chicago Daily News, 195
Chiclero, 122–124
Chile, 8, 173
China, 3, 20, 56, 58, 63, 64, 66, 111, 117, 120, 148, 153, 154, 181–203
China Campaign Committee, 198, 199
Ching-ling, 194
Cholera, 2–4, 13, 14, 16, 17, 151, 156, 166
Cholkar, M., 198
Chongqing, 189, 193, 198
Chung, H. L., 111
Churchill, Winston, 200
Clayton machine, 165
Clegg, Arthur, 199
Coghe, S., 30, 40
Cold War, 12, 15, 113–116
Cole, Gibril R., 79
Colonial medicine, 3, 7, 13, 14, 19, 50, 53–69, 146
Colorni, Umberto, 148, 150, 151, 153, 156
Comintern, 185, 188, 195
Commonwealth, 8
Compans De Brichanteau, Edoardo, 141
Congo, 6, 9, 40n6, 45, 46, 153
Conrad, Sebastian, 3, 4, 10, 15, 17, 18, 142
Contagionism, 3
Convit, Jacinto, 122, 122n28
Corbellini, Gilberto, 149
Correia Mendes, Anníbal Celestino, 33, 40n6, 41
Costa, J., 31
Costa Rica, 118, 122
Cot, Pierre, 185, 186, 186n9, 196, 198
Cotijuba island, 127n32
Coutinho, J. O., 109
Cozzoli, Daniele, 4, 20, 101, 140, 150n5, 152n7, 185
Craddock, S., 15
Cristoforo Colombo battleship, 147, 150
Crozier (then Greenwood), Anna, 1, 2, 5
Cuban missile crisis, 115
Cueto, Marcos, 3, 13, 15, 17, 113, 114, 166
Cundall, Frank, 92
Cuniculus paca paca, 121
Curare, 150, 150n5
Curtin, Philip, 32, 34
Curto, Ramada, 38, 39, 41

Cyrenaica, 13
Czechoslovakia, 185, 187, 200–202

D

da Costa, Luis Neves, 49
Damas Mora, António, 40, 40n6, 41, 44–46, 50
Darwin, John, 5, 8, 17
Dasyprocta azarae, 121
DDT, 110, 110n16, 118, 119
de Almeida, Carlos, 40n6, 45
Deane, Leônidas, 103, 103n3, 109, 110, 110n16, 116, 116n18, 118–121, 120n23, 126
Deane, Maria, 103n3, 109, 110, 110n16, 120n23, 121, 126, 127
de Assunção Velho, João Baptista, 44
De Barros, Juanita, 19, 87, 89, 149
de Brouckère, Louis, 185
DeCordovas, 90
Dedem, W. K. baron van, 66
Dedet, Jean-Pierre, 4, 106, 106n10
de Jong, J. J. P., 56
de Lancastre, António, 38
Delaunay, Robert, 4, 7
Del Boca, Angelo, 140
Deli, 62
de Lisser, Herbert George, 90
de Magalhães, José, 38
Denmark, 8
de Oliveira Peixoto, Cláudio, 125, 125n31
Departamento Nacional de Saúde Publica (DNSP), 170, 171, 173
De Quincey, Thomas, 62, 63
Dernburg, Bernhard, 17
Desowitz, Robert, 107
De, Zhu, 199
d'Harcourt, Joaquin, 184n5
Dias, Jill, 30n1
Dieldrin, 118
Dikötter, F. L., 64
Di Nolfo, Ennio, 8, 9
Diretoria Geral de Saúde Pública (DGSP), 165, 166
Djigounian, A., 117n19
DNERu, *see* National Department of Rural Endemic Diseases
DNSP, *see* Departamento Nacional de Saúde Publica
Dolivet, Louis, 185, 195
Dondo, 44
Dorn, M., 15
Dosyprocta azarae, 120, 121
Douwes, A., 64
Dresden, 202
Dubos, R., 3
Duchêne, Gabrielle, 185
Duclos, Jacques, 188
Dulles, John Foster, 188
Dumett, Raymond E., 78
Duran Jordà, Frederic, 183, 184, 199, 200
Dutch Guiana, 55
Dutta, A. K., 102

E

East Africa, 13, 117, 118
East Indies, 56, 61, 65
Ebro, 201
Echenberg, Myron, 164
Edgerton, David, 8–10, 17
Edler, F., 113
Edmond, R., 2, 14, 58
Egypt, 16, 117, 141, 187, 189, 200
Ehrlich, Paul, 38
Eichengreen, Barry, 8
Eight Route Army, 199
Elder Dempster shipping house, 75, 88, 89
Elliott, John H., 2, 10, 156
El Salvador, 148

Elton, Charles, 169, 170
Emigration Committee, 142
Empires, 1–20, 29–50, 53–69, 86, 88, 89, 140, 144
Engelmann, Lukas, 164
Entebbe, 33, 44
Eritrea, 156
Eritrean colony, 141
Espúndia, 121–123
Ethiopia, 117, 144
Etkind, Alexander, 12
Europe, 3, 11, 12, 14, 16, 32, 34, 55, 58, 60, 145, 149, 163, 164, 166, 168n3, 169, 182, 189, 201, 202
Evandro Chagas Institute, 112, 125

F
Fairchild, Alexander Graham Bell, 118
Falcão, A. R., 118
Fanon, Franz, 30
Fantini, Bernardino, 149
Far East, 153, 154
Farley, John, 13, 78
Farrugia, Peter, 186, 190, 193, 198
Fascism, 12, 140, 185–190, 193, 202
Fay, Richard, 118
Fee, Elisabeth, 114
Ferguson, Niall, 8
Fialho, Amadeu, 170n5, 171, 172
Fielding-Ould, 79, 83
Finlay, A. A. C., 92
Finley, M., 10
Fiocruz, 126
First World War, 11, 67, 140, 182, 184, 185, 189
Fitz-John, J. A., 80
Fleas, 164, 171
Floch, Hervé, 123
Food And Agriculture Organization of the United Nations (FAO), 12, 15
Forattini, Oswaldo P., 120n23, 121
Ford, Jos. C., 92
Foreign Office, 8, 199
Forgacs, David, 156
Foucault, Michel, 54, 55
Fourneau, Ernest, 6, 7
Fradera, Josep Maria, 2, 9, 10
Fraga de Azevedo, João, 37
Framke, Maria, 198
France, 10, 12, 16, 19, 32, 33, 35, 36, 41, 45, 78, 163, 163n2, 166, 186, 188, 196, 200–202
Franco, Francisco, 186, 191, 192, 197
Freetown, 77–87, 89, 95, 96
French Communist Party, 185, 188
French Guyana, 123
French Popular Front, 185
French West Africa, 117
Frenkel, Stephen, 83, 83n7
Friedman, N., 8
Friz, Giuliano, 144, 145
Fry, Margery, 195
Fundação Carlos Chagas, 128
Fyfe, Laurence R, 92

G
Gaastra, F. S., 56
Gabriele, Mariano, 144, 145
Galeano, D., 4
Gallego Vila, Laia, 182
Galli-Valerio, Bruno, 164
Gambia, 79
Gandhi, Mahatma, 185
García Ferrari, M., 4
García Lorca, Fedérico, 185, 191
Garnham, Percy Cyril Claude, 116–121, 117n19, 119n21, 122n27, 123, 125, 127
Gato silvestre, 172
Geddings, Henry, 167

Geison, Gerlad, 3
Geneva, 48, 186, 187, 191, 194, 196
Genoa, 143, 144
Gerbil, 117n19, 120, 120n22
Germanin, 6
Germany, 4, 6, 8, 10, 12, 16, 18, 33, 35, 78, 154, 164, 182, 188, 191, 197
Ghana, 117
Gianturco, Emilio, 150
Gibson, Mary E., 76
Girard, George, 169, 170
Giura, Ludovico, 153, 154
Gleim, Otto, 32
Glucantime, 111
Go, J., 10
Goerg, Odil, 84
Goiter, 112
Gold Standard, 8
Gomes de Rezende, José, 33
Gordon, Sydney, 199
Gorgas Memorial Laboratory, 116
Gorgas, William H., 14, 81
Gradmann, C., 3, 16
Gramsci, Antonio, 141, 156n8
Granollers, 198
Greenhill, R. G., 88
Green-Stewart, Sandria, 96
Greenwood, Anna, *see* Crozier (then Greenwood), Anna
Grífols i Espés, Joan, 184
Grmek, Mirko, 150n5
Grumbach, Solomon, 185
G-strophantin, see Ouabain
Gualandi, Frederico da Costa, 105n6
Guangzhou, 181, 182, 188, 194–197, 200, 201
Guatemala, 122
Guernica, 193
Guerra, Pasquale, 148, 149, 152–154
Guilin, 194
Guiyang, 194
Gussow, Z., 14, 57, 58, 64, 68
Guyana, 122, 123

H

Haapamaki, Michelle Allyson, 197
Hadassah Medical School, 106n9, 124
Hague conventions, 182
Haldane, J. B. S., 197
Hamburg, 6, 202
Hamlin, C., 3
Hamsters, 106n8, 124
Hardy, Anne, 76, 77
Harman, Charles King, 80, 83
Harrison, Mark, 2, 5, 9, 77
Hart, Ansell, 92
Havik, P., 30, 48
Hawaii, 57
Hayer, Alethea, 62
Haynes, D. M., 2, 5
Hayther, 62
Headrick, Daniel, 14, 30
Hebrew University, 106, 124
Hemming, A.W.L., 86
Hemoglobinuric fever, 42
Herrer, Aristides, 120, 120n24
Hertig, Marshall, 116, 116n18, 118
Heyneman, D., 117n19
Hiroshima, 202
Historical Office of the Navy, 144
Hitler, Adolf, 186, 188
Hoare, Arthur, 110n15, 120n22
Hobsbawm, Eric J., 8, 17, 19, 20, 140
Hochman, G., 109, 165, 171
Hochstadt, S., 2
Hoefte, R., 61
Hogan, M. J., 12
Holanda, D., 110
Holman, Brett, 182, 187, 189
Holt, Thomas, 87, 88, 93

Hong Kong, 148, 163, 194
Hoostraal, H., 117n19
Howard–Jones, Norman, 16

I
Ileto, R. C., 17
India, 5, 9, 16, 20, 76, 83, 84n8, 101, 103–105, 105n5, 116n18, 117, 120, 163, 164, 166, 187, 189, 198–200
Indian Medical Service, 76
Indian National Congress, 198
Indian Ocean, 30n1, 148
Indian Plague Commission, 164
Indigenous Medical Assistance Service, 45
Indonesia, 53, 55, 65, 67
Industrial Revolution, 7
Inini, 123
Institut Colonial International, 142
Institute of Preventive Medicine, 112
Instituto Bacteriológico de São Paulo, 102n1
Instituto de Patologia do Norte, 105
The International Cooperation Administration (ICA), 15, 119, 119n20
International Peace Campaign (IPC), 181–203
International Peace Hospital, 199
International Sanitary Conference, 16, 167
Iran, 117, 120
Iraq, 117, 189
Ireland, 2, 8, 10
Istituto Coloniale Italiano (ICI), 141
Italian Benevolent Institute, 150
Italian Navy, 144, 146, 148, 149, 153
Italian Royal Navy, 17, 139–156
Italian Society for Political Economy, 140
Italo-Turkish War, 141
Italy, 2–4, 9, 10, 117, 141, 142n2, 144, 145, 149, 150, 150n5, 152–156, 182, 191, 197

J
Jaguaribe, 110
Jamaica, 3, 75–96, 124, 200
Japan, 8, 64, 182, 190, 193, 195, 197, 200
Java, 14, 56, 61
Jézéquiel, Jules, 185, 194
Jiang, Yizhen, 198
Jin-Cha-Ji, 199
Jogas Jr., Denis Guedes, 102, 102n1, 106n8, 108, 112
Johnson, Ryan, 4, 5
Jones, Alfred, 75–79, 78n1, 81–83, 85–91
Jones, Margaret, 77
Jorge, David, 191
Jorge, Ricardo, 45, 46, 162, 168–170, 168n3, 173, 175
Joseph II, 10
Judson, Peter, 10

K
Kala-azar, 101, 103–105, 107, 111, 112, 117
Kalahari Desert, 169
Kannabateomys amblyonyx amblyonyx, 121
Kellogg Foundation, 113
Kemp, A., 113
Kennedy, John F., 114
Kenya, 116–120
Kerodon rupestris, 171
Khan, Ateeka, 96
Kho-seng Lim, Robert, 192
Killick-Kendrick, R., 106n10, 107

Kingdom of Sardinia, 143–145
King-Harman, R., 80, 83
King Leopold, 153
Kitasato, Shibasaburo, 163
Koch, Robert, 3, 16, 38
Kong Xiangxi, *see* Kung, H. H. Kunming
Koo, Pat, 199
Koo, Wellington, 199
Kopke, Ayres, 33, 38–41, 44, 46
Kotnis, Dwarkanath, 198, 199
Krishna Menon, V. K., 199
Kryukova, A. P., 117n19, 120n22
Kubitschek, Juscelino, 109, 112, 114
Kuhnke, L., 16
Kung, H. H. Kunming, 188
Kuomintang, 188
Kwanza, 45

L

Labanca, Nicola, 140
Lainson, Ralph, 121n25, 125, 126
Lamb, George, 164
Lambert, David, 77
Lampe, P. H. J., 61
Lane, Jane, 175
La Paz, 122
Lapenta, Pedro, 122
La Plata, 144
Latour, Bruno, 15
Latyshev, N. I., 117n19
Lausanne, 196
Laveran, Alphonse, 38
L. braziliensis, *see Leishmania, Leishmania braziliensis*
League of Nations, 38, 44–46, 50, 54, 182, 185, 186, 192, 202
Lebanon, 117
Leffler, Melvin, 12
Leishmania
  Leishmania brasiliensis, 104, 104n4, 107, 108, 120, 121, 123, 124
  Leishmania chagasi, 104, 107, 108
  Leishmania donovani, 103, 107, 111, 121
  Leishmania infantum, 103, 107, 108
  Leishmania tropica, 108, 117n19, 120n22, 121, 124
  Leishmania tropica mexicana, 123
  Leishmania tropica minor, 117n19, 120
Leishmaniasis
  cutaneous leishmaniasis, 101–103, 105, 106, 109, 117, 119–122, 120n24
  tegumentary leishmaniasis, 19, 102, 102n1, 104, 108, 109, 118, 121–123, 125, 126
  visceral leishmaniasis, 101–112, 102n1, 106n8, 107n11, 117–122, 120n23, 126, 127, 127n32
Lemaine, G., 30
Leprosy, 2, 14, 18, 19, 47, 53–69, 122, 122n28
Lesch, J., 150n5
Lester, Alan, 77
Leung, K. C., 58
Leva, Fausto, 144, 147
Lewis, David J., 122n27
Libya, 141
Liebig colon, 10
Li Hua-yu, 197
Lille, 163
Lima Institute of Hygiene and Public Health, 121
Lima, Nisia, 114
Lindemberg, Adolpho Carlos, 102n1
Lisbon, 31, 37, 37n4, 38, 40, 41, 46, 48, 124
Lisbon Geography Society, 34

Lisbon School of Tropical Medicine (LSTM), 31–39, 41, 44, 146
Lisbon Society of Medical Sciences, 32
Liverpool, 4, 78, 79, 83, 86, 89, 95
Liverpool School of Tropical Medicine (LSTM), 19, 34, 76–80, 84, 85, 89, 90, 95, 96
Li Yu-ying, 198
London, 38, 184, 186, 187, 189, 194, 195, 197–202
London School of Hygiene and Tropical Medicine, 34, 116, 124, 125
Long Depression, 140
Long, Edward, 94
Long nineteenth century, 2, 8, 16
Look Lai, Walton, 91
Los Angeles, 148, 196
Louisiana, 57
Lourenço Marques, 44
Louro, Michele L., 185, 188, 189, 192, 201
Löwy, Ilana, 4, 166
LSTM, *see* Lisbon School of Tropical Medicine; Liverpool School of Tropical Medicine
Lu, C. C., 111
Luanda, 32, 33, 38, 44, 45
Lumsden, J., 88
Luna, Simone, 162, 174
Lutz, Adolpho, 119
Lutzomyia
  Lutzomyia intermedia, 105
  Lutzomyia longipalpis, 105, 127n32
  Lutzomyia migonei, 109
  Lutzomyia olmeca, 124n29
  Lutzomyia pessoai, 109
  Lutzomyia whitmani, 109
Luzzati, Alfredo, 148
Lycalopex vetulus, 110
Lynteris, Christos, 164
Lyons, M, 32, 78

M
Macau, 40n6, 45
Macchiavello Varas, Atilio, 173, 174
Macqueen, Adam, 77
Madagascar, 148
Madancy, J. A., 64
Madia, Ernesto, 145, 146
Madrid, 181, 183, 184, 186, 190, 191, 193, 194, 199, 201
Maehle, Andreas-Holger, 150n5
Malaga, 184
Malaria, 4, 13, 15, 19, 32, 47, 75–79, 81, 82, 83n7, 84–87, 89–95, 102, 107, 111–114, 118, 119, 122–124, 148, 149, 154, 156
Malizia, Enrico, 149
Manchuria, 168
Manila, 202
Mann, M., 11
Mannelli, Mario, 148, 153
Manson-Bahr, Philip Edmund Clinton, 116, 116n18, 118
Manson, Patrick, 13, 31, 38, 78, 147
Marapanim, 105
Marco Polo incident, 190
Marsden, Philip Davis, 105n5, 125
Martin, E., 15
Martin, Fredericka, 184n5
Martini, Ferdinando, 156
Martins, Amilcar Vianna, 124
Marx, Karl, 9
Massawa, 153
Mazagan, 151
Mazuy, Madame Rachel, 185, 186, 188, 190
Mazza, Salvador, 120n23
McIntyre, A. R., 150n5
McNeill, J. R., 2
Medical Relief Corps, 192
Medina, Rafael, 122n28

Mediterranean, 4, 16, 103, 104, 106, 107, 117, 120, 122, 153
Meijring, K. H., 66
Mendoza, 106
Menelik, 141
Mercado, Charles E., 90
Mesnil, Félix Étienne Pierre, 121
Mexico, 122, 187, 200
Meyer, Karl, 169
Middle East, 76, 117
Migone Mieres, Luis Enrique, 102
Miguel Bombarda Hospital, 44
Milward, Alan, 12
Minas Gerais, 105n5, 112, 120n23
Ministero degli Affari Esteri, 142
Minsk, 202
Miranda Guedes, 41
Mirzoian, 117n19
Mogadishu, 141, 153
Mojú, 105
Moldavia, 10
Moluccas, 56
Montano, Antonio, 155
Morocco, 148, 153, 182
Moroni, Ercole, 150n5
Moscow, 189
Mosse, Charles, 93
Motta, Mario da Camara, 171
Moulin, Anne-Marie, 4
Mowrer, Edgar E., 195
Mozambique, 38, 41, 43, 44, 48, 50
Mughal Empire, 9
Mukherjee, Debesh, 198
Mumbai, 152, 152n6, 164
Munich, 189, 201
Münzenberg, Willi, 188, 195
Musil, Robert, 11
Mussolini, Benito, 140, 186, 202
Musson, S. P., 92
Myocastor coypus, 121
Myrdal, Gunnar, 114

N
Nagasaki, 202
Nairobi, 116
Namibia, 169
Naples, 149
Nascimento, Dilene Raimundo do, 162, 165
Nation, 1–20, 53, 57, 58, 114, 126, 139–144, 149, 155, 192, 193
National Academy of China, 197
National Academy of Sciences (NAS), 115, 115n17, 116, 116n18, 121
National Department of Rural Endemic Diseases (DNERu), 112, 124
National Institute of Amazonian Research (INPA), 125
National Malaria Service, 109, 111
National Research Council (CNR), 115
Naval hygiene, 145
Naval School, 31, 33
Navarre, Pierre-Just, 147
Needham, Joseph, 197
Nehru, Jawaharlal, 185, 189, 189n18, 192, 198, 199, 201, 202
Neill, Deborah, 5, 6, 19, 30, 149, 183
Neiva, 107n11
Nery-Guimarães, Felipe, 103n3
The Netherlands, 3, 9, 55, 64, 66–68
New England, 10
New York, 47, 196
New Zealand, 57, 200
Nicolle, Pierre, 4
Nigeria, 117
Noel-Baker, Philip, 185
Non-Western medicine, 152
Noroeste do Brasil railroad line, 102
North America, 14, 148, 155
Northern Rhodesia, 43
Northrup, David, 91
Nutrias, 121
Nyasaland, 43

O

Office International d'Hygiène Publique (OIHP), 16, 166–168
Oficina Panamericana de Sanidad, 16
Olivi, Gerardo, 150
Omissi, David E., 182, 183, 185
OPA, *see* Pan American Operation
Opium, 62–67, 69, 153
Opium Wars, 64
Ormsby-Gore, William, 42
Osborne, Michael A., 2, 3, 5, 9
Oslo, 196
Osseo-Asare, A. D., 150n4
Ossorio y Gallardo, Ángel, 190, 191
Osterhammel, Jürgen, 8, 10, 12
Oswaldo Cruz Foundation, 112, 126
Oswaldo Cruz Institute, 103, 103n2, 104, 107, 112, 126
Ouabain, 150, 150n4, 151
Ouabaio, 150

P

Packard, Randall, 14
Padwa, Howard, 63
Pakistan, 117
Palfreeman, Linda, 183
Panama, 116, 118, 121, 121n25, 122
Panama Canal Zone, 112
Pan American Health Conferences, 114
Pan American Health Organization (PAHO), 113, 114, 125, 126, 173
Pan American Operation (OPA), 114
Pan-American Sanitary Bureau (PASB), 13
Pan American Sanitary Conference, 112
Pan American Sanitary Organization, 113
Paquette, Gabriel, 2, 9
Pará, 103, 103n3, 105, 105n7, 125, 127n32, 149
Paraense, L., 106n8
Paraíba, 110, 171, 173
Paramaribo, 60
Paranhos, Ulysses, 102n1
Paris, 4, 16, 38, 186, 188–191, 193, 194, 197–199, 201
Paris Soir, 195
Parreiras, D., 171
Pasquale, Alessandro, 142, 146, 147, 150
Pasteur Institute, 4, 6, 106, 154, 163
Pasteur, Louis, 3, 4, 15, 16, 18
Pasteur Vallery-Radot, Louis, 7
Pati, B., 2
Patton, Adell Jr., 80
Pedroso, Alexandrino M., 120n23
Pemberton, Neil, 163
Penna, Henrique, 103, 104
Pereira Lapa, José, 44
Pernambuco, 112, 171, 173
Peru, 8, 118, 120, 121, 143
Peruvian warts, *see Verruga peruana*
Pessôa, Samuel, 108–110, 120n23, 121, 123, 127
Petersson, Frederick, 188
Petrograd, 7
Philippines, 14, 17
Phlebotomus
  Phlebotomus argentipes, 107
  Phlebotomus caucasicus, 120n22
  Phlebotomus intermedius, 105, 106n8, 107
  Phlebotomus longipalpis, 105, 106n8, 107n11, 110n16, 111, 118
  Phlebotomus migonei, 109
  Phlebotomus papatasi, 106, 107
Phlebotomus pessoai, 109
  Phlebotomus whitmani, 108

Phthisis, 155
Phyllomys, 105
Pian-bois, 122
Piancó river, 110
Picasso, Pablo, 193
Piccolino, Giulia, 153
Piemonte, Royal Navy ship, 154
Pifano, Felix, 120n23
Pilsen, 196
Pinotti, Mário, 109
Piratuba, 105
Pires-Alves, F., 113
Plague, 2, 3, 13, 16, 20, 104, 152, 161–176
Plague, sylvatic, 20, 104, 162, 163, 167–176
Poland, 187, 202
Pontiac, Peter, 67, 68
Pope, J. W., 7
Portugal, 3, 8, 30, 30n1, 31, 34, 35, 37n4, 38, 44, 45, 47, 48, 50, 146, 168
Portuguese Guinea, 48
Pottevin, Henri, 166, 168
Pouderoux, Paul, 185, 187
Povalishina, T. P., 110n15
Power, Helen J., 77
Prado Barrientos, Luiz, 122
Prague, 196
Prates, Manuel Máximo, 44
Príncipe, 39–43, 45, 50
Pronk, J. B., 67
Prout, William, 76, 77, 79–82, 82n5, 84–87, 89–96, 91n12
Providencia island, 122
Prudhommeaux, Jules, 190n19, 202n47, 202n48
Prussia, 3
Puerto Rico, 112
Pulex cheopis, 164
Punjab, 164
Punta Arenas, 152

Q
Quarantine, 3, 7, 15, 80, 161
Quarantinism, 16
Quinian, Sean, 58

R
Rajchman, Ludwig, 45
Rangel, Elizabeth, 106n8, 126
Rangoon, 202
Rasmussen, A., 16
Rassemblement Universel pour la Paix (RUP), *see* International Peace Campaign
Rat
  black rat, 163n2
  brown rat, 163n2
  Rattus norvegicus (*see* Rat, brown rat)
  Rattus rattus (*see* Rat, black rat)
Ratoeiros, 165, 166
Recife, 112
Red Cross, 183, 191, 192, 197
Ren, Ke, 185, 190, 193
René Rachou Research Center, 112
Republic of Genoa, 143
Rho, Filippo, 143, 145–147, 152–154
Rhombomys opimus, 120
Rhône-Poulenc, 6, 111
Ribbentrop-Molotov, pact, 189
Ribeirão Preto (SP), 112
Ricciardi, Toni, 140
Rio de Janeiro, 4, 112, 124, 142, 162, 165–167, 171, 176
Rio de la Plata, 144
Rockefeller Foundation, 13, 15, 47, 103n2, 115, 175
Rodrigo, Antonina, 185, 187
Romaña, Cecílio, 103n2, 106, 120n23
Romero, Jesus, 122n28
Rongzhen, Nie, 199
Roosevelt, Franklin D., 200

Roosevelt Hospital, 148
Roselli, Francesco, 148
Rosenberg, C. E., 54
Ross, Ronald, 76–79, 79n2, 81–83, 83n6, 83n7, 84n8, 85, 92, 95, 96
Rotterdam, 202
Roux, Émile, 16
Roxburgh, T. L., 92
Royal Office of Statistics, 142
Royal Society, 83n6, 83n7
Rude, Maxine, 119
Rumania, 187
RUP/IPC, *see* International Peace Campaign
Rush, J. R., 66
Russia, 7, 14, 117, 117n19
Ryall, Lucy, 143, 144

S

Sabbe-Clarke, E. E., 32
Sabin, Albert B., 115, 116
Sahara, 117
Saigon, 153
Salazar, Antonio, 9, 47–48
Salvador de Bahia, 112
Salvadori, Massimo, 145
San Andrés University, 122
Sanarelli, Giuseppe, 13
San Diego, 148
San Gottardo ship, 148
Sanitation, 2, 3, 15, 40, 42, 44, 80, 86, 95, 153
Sanitationism, 16
San Raffaele Society, 150
Santa Catarina, 119
Sant'Ana, Firmino, 43, 44
Santos, 148, 149, 164
São Paulo, 102n1, 108, 112, 121, 140, 141, 161, 162, 164, 165
São Paulo School of Medicine, 108
São Paulo State Department of Health, 108
São Tomé, 40–42, 45
Sartori, A., 9
Savoy, House of, 143
Scarfone, Marianna, 156
Schauff, Frank, 188
Schiebinger, Londa, 58
Schilling, Godfried W, 59, 60
Schmidt, Jammy, 202n48
Schockov, Yakimoff, 117n19
Schwartz, A. J., 8
Scize, Pierre, 195
Scotland, 10, 14
Scudder, John, 200
Second World War, 55, 67, 189, 200, 202
Secord, Jim, 19
Sedan, battle of, 16
Senegal, 4
Sen Gupta, P. C., 116n18
Sequeira, Fontoura de, 44
Sergent, Edmond, 106
Sergent, Étienne, 106
Serviço Nacional de Peste, 162
Serviço Permanente de Prevenção e Combate à Peste Bubônica no Sul de Angola, 169
Seychelles Islands, 148, 153
Shanghai, 194, 195
Shanxi- Chahar-Hebei, 199
Shapiro, Frederick, 30
Shaw, Jeffrey, 106n8, 125, 126
Shekhanov, M. V., 120n22
Shortt, Henry Edward, 106n9, 107
Siberian marmot, 168
Sid-ben-Nur, 151
Sierra Leone, 3
Sila, 156
Silva, Francisco Venáncio da, 169
Silva Junior, Marcelo, 171, 172, 175
Silva, Kalinga Tudor, 76
Silva, Matheus Alves Duarte da, 161–176
Silveira, Fernando Tobias, 127n32

Simon, Roland, 174, 175
Simond, Paul-Louis, 164, 166
Simons, C., 61
Sinclair, A. C., 92
Singapore, 202
Slater, Leo, 14
Sleeping sickness, 6, 13, 18, 31–33, 38, 39, 41–46, 48–50, 76, 102
Smallpox, 16, 80, 165
Smith, Clarence, 41
Smith, Philip, 150n5
Snelders, Stephen, 18, 55, 59–62, 64–67
Snowden, Frank, 149
Social Darwinism, 57
Society for the Protection of the Italian Immigrants, 150
Society of Suriname, 55
Solé i Sabaté, Josep Maria, 182, 191, 200
Solomon Gross, S., 7
Soluri, John, 92
Solustibosan, 111
Somalia, 9, 140, 144, 153, 156
Somaliland, 189
Soong, Ching-ling, 194
Soong, T. V., 194, 201
Soper, Fred L., 104, 114, 175
Soure, 105
South-Africa, 68, 172, 200
South America, 3, 4, 10, 12–14, 16, 18–20, 55, 122, 140, 141, 143, 144, 147–149, 155, 174
Soviet Central Asia, 120
Soviet Union, *see* USSR
Spain, 3, 9, 20, 181–203
Spanish-American War, 3, 145
Spanish Civil War, 183, 190, 201
Spanish International Brigades, 199
Spanish Republic, 188, 191, 197
Special Public Health Service(SESP), 112, 113
Spitzer, Leo, 79, 84
Sri Lanka, 76
Staffetta, ship, 148
Standard Oil, 12
Stepan, Nancy, 161
Steverding, D., 32
Stewart, Agnes, 103n3
Stewart, R. J., 184n5
Stewart, Roderick, 199
Stewart, Sharon, 199
Stockholm, 196
Strang, D., 49
Strangways-Dixon, John, 124, 125
Subaltern, 20, 141, 142, 146, 153, 155, 156
Sudan, 117, 118, 182, 189
Sullivan, R. J., 17
Sundy Farm, 42
Suriname, 53, 55–56, 58–62, 66–68
Suvorova, L. G., 120n22
Switzerland, 140, 200
Syria, 117, 188, 189
Szasz, Thomas, 55

T

Tagore, Rabindranath, 185
Talamonti, Luigi, 141
Tangier, 151
Tashkent, 111
Taylor, A. J. P., 10, 11, 17
Taylor, Logan M., 80–82, 82n5
Teixeira de Sousa, Antonio, 36
Teixeira, Luiz Antonio, 161
Teles, Silva, 38
Tete, 43
Theodor, Oskar, 106, 107
Thomas, Bryn K., 150n5
Thrichomys apereoides, 171
Tilley, Helen, 30, 39
Times, the, 6, 80
Timor, 40n6, 45
Tokyo, 192, 192n22, 202
Topiwala, Harshad, 2

Toxoplasmosis, 116
Trachoma, 13, 14, 112
Tracy, G. S., 14
Transfusions, 183, 184, 200
Trimarchi, Giuseppe, 156
Trinidad and Tobago, 119
Tripolitania, 13
Tropical medicine, 1–3, 5–7, 9, 10, 13–15, 18, 19, 29–50, 102, 104, 113–116, 125, 139–156, 161, 175, 183
Trueta, Josep, 184, 185, 187, 199
Trypanosoma, 32
Trypanosomiases, 116
Trypanosomiasis, 6, 7, 47, 48, 50, 103
Tuberculosis, 149, 152, 152n7, 155
Tunisia, 141
Turkey, 64
Turkistan, 117
Turkmenistan, 117n19, 120, 120n22
Typhus, 4, 14

U
UNESCO, 12, 15
United Fruit Company, 12, 87, 88
United Kingdom of Great Britain and Northern Ireland, 8
United Nations Organization, 12
United States of America (USA), 57, 62, 88, 113–116, 145, 150, 155, 169, 186, 187, 196, 200, 201, 203
Universal Congress for Peace, 186, 191
Universidad Nacional Autónoma de Mexico, 122
University of Brasília, 125
University of Naples, 142, 143, 145, 146
University of São Paulo, 108, 116, 121
Uruguay, 115
USA, *see* United States of America
US Agency for International Development (USAID), 15, 119n20
USSR, 12, 15, 114, 187, 188, 193, 202
Uta, 118, 120, 120n24, 122, 123
Uttar Pradesh, 198
Uzbek Soviet Socialist Republic, 111

V
Valencia, 193, 194
Valéry, Paul, 185
van Bergen, L., 62
van den Doel, H. W., 56
van Doorn, J. A. A., 56
van Driel, B. M., 62
van Hasselaar, A., 60
van Luijk, E. V. V., 66
van Ours, J. C., 66
Vanvugt, E., 66
Vargas, Getúlio, 109
Vasconcellos Coelho, Marcello de, 124
Vasconcellos, Henrique Figueiredo de, 165, 166
Velmet, Aro, 76, 77
Venezuela, 120n23, 122
Venice International Convention, 16
Verruga peruana, 118
Vianna, Gaspar, 104, 104n4, 105, 105n5, 121
Vianna Martins, Amilcar, 124
Villaggio Duca degli Abruzzi, 141
Villarroya, Juan, 182, 191, 200
Villela, Arnaldo José, 41
von Humboldt, Alexander, 150
von Paumgartten Deane, Maria, 110
Vyce, Mandy, 96

W
Wagner, Florian, 142
Wales, 10
Wallerstein, Immanuel, 5, 6
Warsaw, 202
Webb, Jr., James L. A., 76
Weindling, Paul, 4, 7, 14, 15
Weiss, Pedro, 121, 122
Wellcome Trust, 125
West Bengal, 118
Western, 2, 17, 19, 57, 64, 150–152, 154, 187, 191, 193, 197
Western medicine, 16, 17, 101
West Indies, 18, 53–69, 88, 124
Wilkinson, Lise, 77
Will, H. A., 88
Winslow, Amory, 114
Winterbottom, Thomas, 32
Wise, K. S., 76
Worboys, Michael, 16, 30, 77, 78
World Expo, 193
World Health Assembly, 114, 126
World Health Organization (WHO), 12–15, 46–48, 50, 101, 113, 114, 116, 118, 122n26, 124–126
World War II, 46, 54, 55, 67, 113, 189, 200, 202
Wright, Willard H., 57, 116n18, 121
Wuhan, 181, 193, 194, 198, 201
Wutai, 199
Wyllie, J., 42

X
Xingzhi, Tao, 185

Y
Yakimoff, W. L, 117n19
Yan'an, 198
Yangfu, Zeng, 182
Yatsen, Sun, 194
Yaws, 112
Yellow fever, 2, 4, 13, 16, 76, 81, 103, 104, 111, 112, 114, 148, 149, 165, 166, 174, 175, 183
Yellow Fever Service, 103
Yersin, Alexandre, 163, 164
Yersinia pestis, 163, 168
Yucatan, 122
Yuxiang, Feng, 188

Z
Zambezi, 43
Zambezia, 43
Zanzibar, 147
Zedong, Mao, 199
Zhabei, 194

FSC
www.fsc.org
MIX
Papier | Fördert
gute Waldnutzung
FSC® C083411

Kolibri 360 GmbH
Ferdinand-Jühlke-Straße 7
99095 Erfurt, Deutschland
produktsicherheit@kolibri360.de